Induced Innovation and Productivity-Enhancing, Resource-Conserving Technologies in Central America

European University Studies
Europäische Hochschulschriften
Publications Universitaires Européennes

Series V
Economics and Management

Reihe V Série V
Volks- und Betriebswirtschaft
Sciences économiques, gestion d'entreprise

Vol./Bd. 3041

PETER LANG
Frankfurt am Main · Berlin · Bern · Bruxelles · New York · Oxford · Wien

Monika Barbara Zurek

Induced Innovation and Productivity-Enhancing, Resource-Conserving Technologies in Central America

The Supply of Soil Conservation Practices and Small-Scale Farmers' Adoption in Guatemala and El Salvador

PETER LANG
Europäischer Verlag der Wissenschaften

Bibliographic Information published by Die Deutsche Bibliothek
Die Deutsche Bibliothek lists this publication in the Deutsche Nationalbibliografie; detailed bibliographic data is available in the internet at <http://dnb.ddb.de>.

Zugl.: Gießen, Univ., Diss., 2002

D 26
ISSN 0531-7339
ISBN 3-631-51970-2
US-ISBN 0-8204-6580-1

www.peterlang.de

Table of Contents

List of abbreviations

CA	Central America
CENTA	Centro Nacional de Tecnología Agropecuaria y Forestal (National Agricultural Research Organization of El Salvador)
CIMMYT	Centro Internacional de Mejoramiento de Maiz y Trigo (International Maize and Wheat Improvement Center)
FAO	Food and Agriculture Organization of the United Nations
GNP	Gross National Product
ha	Hectare (1 Hectare = 10,000 m^2)
ICTA	Instituto de Ciencias y Tecnología Agrícolas (Guatemalan National Agricultural Research Organization)
ISRIC	International Soil Reference and Information Centre
inh./km^2	Inhabitants/Sqare kilometer
kg	Kilogram
mz	manzana (1 manzana = 0.7 ha)
N	Nitrogen
NARS	National Agricultural Research System
n.d.	no data available
NGO	Non – Governmental Organisation
NRM	Natural Resource Management
PERC	Productivity enhancing and resource conserving
PRM	Programa Regional de Maíz (Regional Maize Program)
UNEP	United Nations Environment Program
USLE	Universal Soil Loss Equation

Acknowledgements

Conducting my fieldwork and writing this thesis would not have been possible without the help and encouragement of a large number of people, to all of whom I am deeply indebted.

First I must thank my thesis advisor Prof. Dr. Ernst-August Nuppenau for accepting me as a PhD student, for his long-lasting support, and for helpful and constructive discussions during the write-up of the thesis and my time in Giessen. I would also very much like to thank my second thesis advisor, Prof. Dr. Gregory Traxler, for his constant support in times of doubt, for fascinating discussions about concepts and theories and their application to this research, and for his very helpful reviews of the first versions of the thesis.

I am especially grateful to CIMMYT - the International Maize and Wheat Improvement Center - for accepting me as a pre-doctoral fellow in the Economics Program and for providing me with the necessary infrastructure and logistical support to carry out my research. Dr. Gustavo Sain, based at CIMMYT's Regional Office for Central America and the Caribbean in San José, Costa Rica, was my original supervisor and without him none of my work in Central America would have been possible. He introduced me to the region and the CIMMYT collaborators, had great patience in answering my never-ending questions, helped me to set up the fieldwork, constantly discussed ideas and new concepts and supported me with his generous friendship. In addition, I am very grateful to Marlen Montoya, Carlos Bonilla and Andres Jauregui from the CIMMYT Costa Rica office for their friendship and support during my three years working with CIMMYT in San José.

Also in Central America, I am particularly grateful to the farmers who sat through long discussions and answered - with great patience - my questions during the surveys in Guatemala and El Salvador. Without their cooperation, none of my field research would have been possible!

In Guatemala, Jérôme Fournier, at the time a PhD student in the CIMMYT Guatemala office, introduced me to the Polochic Valley. I am very grateful for his support in setting up my field research in the area, for sharing many of his field experiences with me, and for the fun-times we had while learning more about 'life in the tropics'. Dr. Jorge Bolaños, leader of the CIMMYT Guatemala office, provided vibrant discussions and insights about rural-development issues in Central America, and I would like to thank him for his support during my time in the region. Furthermore, I am grateful to Miriam Hernádez, Wiliam Quemé, Mariela González and Don Landelino Aguirre of the CIMMYT Guatemala office for all their help. In the Polochic Valley, I am especially indebted to Leonel Chavez, without whose explanations, help and great organizational skills, the work in the valley would have been so much more difficult and whose family always made me feel very welcome. In addition, Otoniel Gracia from the ICTA office in Panzós provided a lot of help during the fieldwork. And I would also like to thank all the surveyors from the valley who demonstrated great patience in collecting the information I needed.

I am very grateful for the help I received from the members of the *Programa Regional de Maíz* in Central America. They supported me during field visits and never failed to help me learn more about the problems and important issues of the region. My special thanks goes to Cristina Choto de Cerna, who was a great mentor for learning about El Salvador and who helped to organize the field survey and the surveyors in Nueva Concepción. And who was great fun to work with!

I would like to thank my current supervisor Dr. Prabhu Pingali, former director of the CIMMYT Economics Program and currently director of the Agricultural and Development Economics Division at FAO, for his support during my time in Costa Rica and for allowing me the time to finish this thesis while I was simultaneously working with him on a new project. In addition, I am grateful to the other members of the CIMMYT Economics Program in Mexico, with whom I could always discuss ideas and fieldwork results. In CIMMYT headquarters, I would also like to thank Dagoberto Flores, who provided invaluable help for the farmers' surveys, Jens Riis-Jacobsen and Lone Badstue for their friendship despite having to comment on the first versions of this thesis, and Janin Trinidad for her invaluable help through the last meters before the finishing line. In addition, I am grateful to John Woolston, whose good conversations and cups of teas helped me to get through long weekend working-hours, and to all the other people in HQ that made my life at CIMMYT so interesting and pleasant.

I am happy to acknowledge the German Federal Ministry of Economic Cooperation and Development (BMZ) for its financial support by assigning me to work within the project "Accelerating the adoption of productivity-enhancing, resource-conserving (PERC) technologies in maize-based cropping systems in Central America", which was funded by the Ministry and executed by CIMMYT.

Last but not least, I would like to say an especially big 'Thank you' to my parents, Prof. Dr. Ernst Zurek and Dr. Barbara Zurek, my sister Christina Zurek and my friend Randolph Watpool. During all these years of work, they never let me down; they provided unfailing support, read and re-read all the different versions of the thesis, discussed all my questions and doubts, and gave me the strength and confidence to finish. THANK YOU!

Mexico City, October 2002

1 Introduction

The overuse of natural resources combined with their pollution resulting from human activities has developed into one of the most pressing problems worldwide. The concept of sustainability has emerged in recent decades to address this issue and to provide solutions that ensure the survival of natural and human systems in the future. This paradigm consists of various elements, which try to combine ecological and socio-economic factors. Nevertheless, the diversity of elements and the wide variation of local conditions complicate its practical implementation. Experiences gained from implementation attempts in recent years demonstrate that the success of solutions in one location does not guarantee its transferability to other locations. Understanding the factors that govern technical change directed towards a more sustainable use of natural resources is therefore important for making the concept work.

Agriculture plays a decisive role in the demand and use of natural resources. Thus, technical change in this sector will be crucial for the successful implementation of the sustainability concept. For agricultural production the careful management of fragile soil resources is the very basis for its long-term success. Nevertheless, mismanagement of this resource is common in many places. Accelerated wind and water erosion rates are the most visible outcome of this mismanagement, leading to a constant threat of soil fertility and with this agricultural production. Central America is a good example in this respect, as in particular water erosion is one of the most threatening problems for agriculture in the region.

In Central America accelerated soil erosion rates, mainly due to water erosion, are a very common problem in many agricultural areas. It is even estimated that up to 80% of the agricultural land show some signs of soil degradation due to human activity. Agricultural mismanagement and deforestation, followed by overexploitation and overgrazing, have been identified as the most important factors contributing to the problem (GLASOD study, ISRIC/INEP 1991). Two factors mainly influence the occurrence and the rates of soil erosion: the agro-ecological characteristics and the land use patterns prevalent in the region.

Worldwide there exist many efforts to mitigate the existing soil erosion and degradation problems. Also in Central America the effects of the high erosion rates and the resulting soil fertility problems have been a long-lasting concern for governments, foreign aid agencies, and civil society groups. But despite many efforts by a large number of regional and national projects and programs to raise awareness about the problem and develop and promote different new agricultural practices and soil conservation technologies, overall adoption levels of these environmental innovations have been relatively low. Resource constraints, the resulting subsistence production, and weak institutional settings in many of these nations lead to high opportunity costs for resource conservation. This is especially the case for small-scale farmers, who are the majority of land users in the most fragile and marginal areas.

1.1 Focus and objectives of the study

The study contributes to an overall understanding of the supply and demand for environmental innovations. Small-scale farmers put short-term productivity considerations before concerns for conservation of their natural resource base. Often farmers do not see short-term benefits from the use of pure conservation techniques or incur additional costs even though from society's perspective the adoption of conservation technologies can have large benefits. But for adoption to occur it is important to understand the perceived benefits of environmental innovations for farmers.

The "Induced Innovation Theory of Technical and Institutional Change", developed by Y. Hayami and V. Ruttan, is used to help understand the wider context in which technical change in agriculture and natural resource management takes place. This theory explains the development and adoption of new practices by relating the choices of farmers to their perception of resource scarcities. If new technologies diminish these scarcities farmers are more likely to adopt the new practice. Technologies that are not in consort with factor endowments will be unattractive to potential adopters. This theory can serve to explain the development and adoption of environmental innovations, which differ in various aspects from commercial innovations.

Two case studies are used to illustrate the importance of both demand and supply of technologies: Demand by farmers for environmental technologies and the factors that govern their adoption are illustrated with data from the Polochic Valley, Guatemala. This first case study looks at the reasons for the adoption of a successful conservation technology (legume intercropping system). Farmers use the technique, for profitability as well as for conservation motives. On the supply side mechanisms must be in place to ensure the continuous development of practices that correspond to farmers' resource scarcities. The second case study, which uses data from the county of Nuveva Concepción in El Salvador, examines the extent to which technology suppliers are aware of characteristics of conservation technologies. Differences in perspectives on resource degradation and the benefits of conservation practices between technology promoters and users can lead to the development and promotion of practices that do not correspond to farmers circumstances, resulting in low adoption levels.

The author likes to contribute with the presented study to the overall discussion of which paths and mechanisms to pursue in the development of technologies that are aimed at ameliorating the increasingly pressing resource degradation problems worldwide.

1.2 Thesis outline

The study presented here first looks at land degradation and its associated problems in Central America. Agro-ecological characteristics and land use patterns of the region and their connection with the degradation problem are described in Chapter 2. In Chapter 3 the notion of technical change and environmental innovations is explored in detail. The development and implementation of this type of new technologies is investigated in the context of the "Theory of Induced Innovation", developed by Y. Hayami and V. Ruttan. In the following two chapters two case studies from Central America are presented. In Chapter 4 the factors influencing the adoption of a productivity-enhancing, resource-conserving (PERC) technology in the Central American setting, the use of the legume Mucuna as a cover crop, and the linkages among these factors are presented in a case study from Guatemala. The supply of conservation practices to farmers, the selection process leading to the choice offered and the organizations working in this field are studied in a second case study from El Salvador. In Chapter 6 final conclusions and lessons learned are presented.

2 Soil Erosion in Central America – Main Causes and Promoted Solutions

For decades soil erosion has been considered a serious problem in farming systems of the tropics and subtropics. Central America is no exception. The reasons for the occurrence of the problem and its possible solutions vary among regions.

In the following chapter background information is provided on soil erosion and the problems associated with soil degradation. Hypotheses are developed on factors influencing the extent of soil erosion in Central America and solutions to combat the erosion problem, which are offered to farmers in the region, are studied. In addition, a classification of soil conservation technologies utilized in Central America is introduced. This information provides the context for the two case studies presented in the later chapters on selection and adoption of conservation practices in Guatemala and El Salvador.

2.1 Soil degradation and soil erosion

Within the ecosphere soils fulfill some very important functions, which include the provisioning of habitats for microorganisms, animals, and plants and the regulation of nutrient, mineral, water, and air cycling. For man the use of soils for the production of food, fiber, fuel, and feed is of utmost importance (Scheffer and Schachtschnabel 1998). Soils can be distinguished by origin, chemical composition, texture, and depth, which determine their ability to support plant development. This ability is called soil fertility (Scheffer and Schachtschnabel 1998) and shapes the production potential of different soils.

2.1.1 Soil degradation and soil erosion worldwide

Soils can deteriorate through natural processes and/or human activity. Soil degradation can be defined as the diminution or complete loss of the productive potential of the soil for current and/or future use (Blaikie and Brookfield 1987). The following processes can cause soil degradation (Grohs 1994, Wild 1995):

- Erosion caused by water and/or wind
- Chemical or physical changes in soil properties
- Excess of salts

It has been estimated that of the total land area of this planet less than 25% is suited for agricultural use (El-Swaify 1994). Half of the potentially agricultural land is today taken into production, while the other half lies under forests and grasslands and can be utilized only with constraints. It is estimated that every five minutes about 10 ha of agricultural land are lost due to its utilization. Of these 10 ha five degrade because of soil erosion, three by salinization, one through other processes and one due to urbanization (El-Swaify 1994). In the only study so far that assesses soil degradation on a global scale (GLASOD Study), carried out by the International Soil Reference and Information Centre (ISRIC) for UNEP at the end of the 1980s, it was calculated that 23% of all agricultural land is affected by degradation (38% of all crop land, 21 % of permanent pastures, 18% of forests). Of the degraded area 16% are damaged so strongly that they cannot be used any longer for farming (Wood *et al.* 2000).

Figure 2-1: Reasons for human-induced soil degradation and their relevance in the different regions of the world

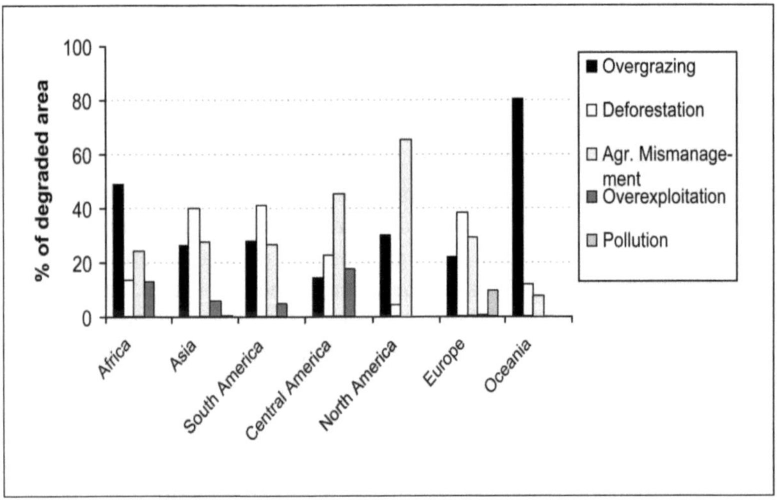

Source: ISRIC/UNEP 1991

The GLASOD study investigates the main human activities, which lead to one of the processes causing soil degradation. The following activities are named in the study (Figure 2-1):

- Overgrazing: Overgrazing is worldwide the most important cause of soil erosion. It is especially severe in Africa and Oceania and the second most important reason in North America.

- Deforestation: The second most important cause worldwide. Asia, South America and Europe are especially affected, while it is the second most important reason in Central America.
- Agricultural Mismanagement: Agricultural Mismanagement plays the most important role in Central and North America, but it is reason number two in Africa, Asia, South America and Europe.
- Overexploitation: Overexploitation only plays a small part in the overall erosion problem.
- Pollution: Only very few soils are degraded due to pollution and so far it has only become a problem in Europe.

The study also states that erosion caused by water is the main reason for soil degradation on all continents and is estimated to affect 50% of the world's land area (Figure 2-2). Water erosion is also named as the factor causing the degradation of 75% of the soils classified as strongly deteriorated (ISRIC/UNEP 1991). In addition, Lal (1989, cited in Napier and Sommers 1993) estimates that 35% of the world's land area is seriously affected by soil erosion. About 44% of agriculturally used land in Asia, 60% in Australia and 40% in Africa are classified as degraded due to erosion measured as a loss in soil fertility (Napier and Sommers 1993).

Soil erosion leads to the "physical removal of fertile topsoil" (Lal 1983) as the process of removing the upper layer of the soil surface by either wind and/or water activity can be defined. The soil lost on one site can then be transported to another site and deposited there. In this thesis only soil loss due to water is investigated, as this is the predominant form of soil erosion occurring in Central America. Thus the terms "soil erosion" or "erosion" in the following only refer to water erosion, if not stated otherwise.

El-Swaify *et al.* (1983, cited in Grohs 1994) define water erosion as "the wearing away of the land surface by the action of water as the geological agent". It can be found in different forms: Interrill erosion, rill erosion, gully erosion, land slides or mass flow and stream bank or coastal erosion (Lal 1983), which differ in the amount of soil moved. Especially the first two forms of water erosion are in many cases slow and difficult to detect. This makes it hard for farmers to see the problem, especially in its early stage, leading to the underestimation of its extent (Rickson et al. 1993).

2.1.2 Factors influencing soil erosion

Different models have been developed to estimate the amount of soil loss through erosion. They include the measurement of tracer movements in the soil

Figure 2-2: Map of soil degradation worldwide

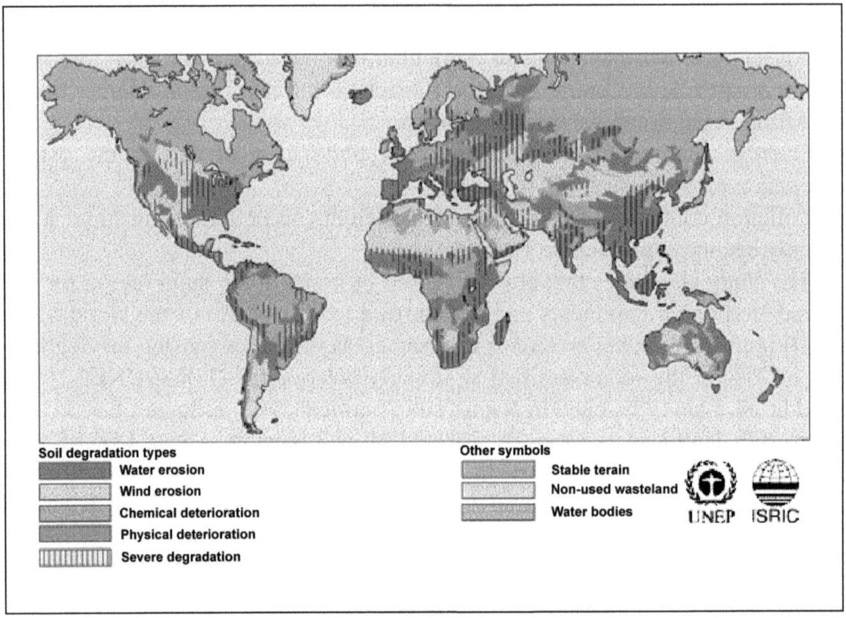

Source: ISRIC/UNEP 1991

over time or the simulation of rainfall events together with the measurement of lost soil. The oldest model, which is still most widely used under field conditions, is the so-called Universal Soil Loss Equation (USLE) (Wild 1995, Scheffer and Schachtschnabel 1998):

$$L = R * K * L * S * C * P$$

L = Average soil loss, R = Erosivity of rainfall, K = Erodibility of the soil, L = Slope length, S = Slope inclination, C = Soil cover due to natural vegetation and cropping system, P = Soil protective measures

In the USLE the most important factors causing soil erosion are named. They include natural factors, like rainfall intensity (R), topography (L and S), soil type (K), and soil cover due to vegetation (C). Here it can be seen that soil erosion is a natural process that can occur in all environments. Soil erosion rates are usually higher in the tropical and sub-tropical regions of the world than in the temperate zones. While water erosion is more frequent in the humid tropics due to the occurrence of high intensity rainfalls, wind erosion plays a more important role in semi-arid and arid regions (Wild 1995). Differences in observed erosion rates between temperate and tropical regions can be explained with a higher fragility of the more weathered soils in the tropics in comparison to the more robust soils in the temperate climates (Anderson and Thampapillai 1990, El-Swaify 1994). Their higher susceptibility to erosion (Lal 1984, cited in Erenstein 1999) together with extreme climatic events leads to on average higher amounts of soil loss (Grohs 1994).

Human activities influence soil erosion rates by altering the vegetative cover of the soil while using certain cropping practices or by implementing soil protective measures. In the USLE this is expressed in factors C and P. In many cases human activities lead to soil erosion rates higher than under natural conditions. Part of agricultural and forestry activities is the removal of the natural vegetation for crop planting or because forests are cut down as a source of raw material. This leaves the soil exposed to the impact of rain and wind and leads to high erosion rates (Napier and Sommers 1993, Wild 1995, Scheffer and Schachtschnalbel 1998).

2.1.3 Effects of soil erosion

Soil erosion can lead to on-site and off-site effects. On-site effects are directly incurred at the site, e.g. the field, where erosion takes place, while off-site effects occur outside the actual erosion zone.

Soil erosion reduces soil depth and affects physical and chemical properties of the soil by changing the aggregate structure and the water, air, and nutrient cycles within the soil. These so-called on-site effects can lead to (Lal 1987, cited in Grohs 1994):

- The leaching of nutrients.
- The diminution of rooting depth for plants.
- The loss of soil organic matter.
- The reduction in plant-available water.

This in return has a direct impact on plant growth, which can be also described as adversely affecting soil fertility in the short- and in the long-run. This decline then directly threatens farm productivity, putting particularly resource-poor farm households in developing countries at risk as they mainly rely on agriculture as their source of income. On a national level, this reduction in farm output can endanger the development of economies dependant on agricultural products for export earnings, which is the case with most developing nations (Lutz *et al.* 1994). In addition, it can result in the need for an increased spending on food, feed, or fiber imports from other countries. This then leads to the use of financial resources necessary for domestic investments to cover the unavoidable import costs (Napier and Sommers 1993).

Nevertheless, it is quite difficult to quantify erosion effects on soil fertility reduction. As plant growth depends on the complex interactions between soil and plant characteristics and the climate it is almost impossible to establish a mono-causal relationship between the erosion rate, related soil degradation, and the decline in crop output (Lal 1983). Several studies have looked into this relationship in more detail. As can be seen in Table 2-1 the percentage of yield

Table 2-1: Interaction between soil erosion and decline in soil fertility

Crop	Percent yield loss per cm of lost soil	Country	Source
Maize	2.0 – 3.4	USA	Lyles 1975
Maize	2.0	USA	Ruthenberg 1980
Maize	2.7 – 4.2	Sierra Leone	Biot 1989
Wheat	0.8 – 3.7	USA	Lyles 1975
Wheat	0.5		Schröder 1974
Sorghum	1.6 – 2.2	USA	Lyles 1975
Sorghum	8.0	India	Vittal 1990
Cowpea	3.3 – 4.1	Sierra Leone	Biot 1989

Source: Grohs 1994

reduction per centimeter of lost soil varies not only between crops but also within crops, and the yield loss estimates also differ for the same crop in different locations. High rates of soil loss in a deep soil of volcanic origin with a high production potential might not result in any reduction in yield for a long time, while already a small amount of lost soil on a shallow, poor soil is likely to lead quickly to a severe output decline (Barbier 1988, Biot and Xi 1993). Figure 2-3 demonstrates both cases: Soil A is a rather poor soil, while soil B represents the fertile, volcanic soil. In soil A yields decline much faster than in B. Therefore the assessment of erosion effects can vary a lot between locations.

Figure 2-3: Hypothetical relationship between cumulative soil erosion and yield for two different soil types

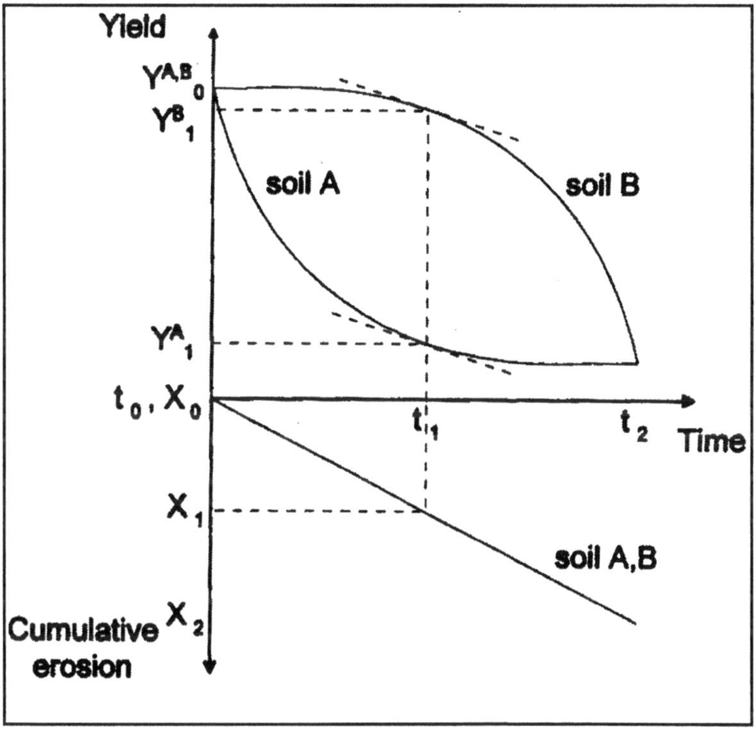

Source: Erenstein 1999

Figure 2-3 furthermore demonstrates another problem associated with assessing the relationship between erosion and soil fertility loss: This relationship is very likely not a linear one. Therefore, depending on soil type, productivity might decline with a sharp drop and then only change very little or vice versa (Biot and Xi 1993). And, as Erenstein (1999) calls it, "most current agricultural land has already been subject to past erosion", which further complicates the understanding of the relationship between degradation and soil fertility decline. Soil loss and its associated damage accumulate over time. So even if the amount lost in one year is relatively small, it can add to the already existing effects of erosion (Pagiola 1994). Depending over which time period erosion and crop production are measured, or when soil conservation measures are introduced, the implications of results can be quite different. For a soil that has lost most of its soil fertility the implementation of erosion control measures might reduce the amount of soil loss but will only have a marginal effect on crop production. In contrast in a soil that is still relatively intact conservation measures might save the soil productive potential substantially (Erenstein 1999).

This discussion leads to another problem that will be investigated later in more detail: the economic valuation of soil fertility loss and soil conservation. For farmers declining soil productive potential only translates into an economic problem if crop outputs are affected. Thus if erosion occurs but its effects are not felt by the farmer through a decline in yields, the costs of implementing soil conservation practices, which are usually associated with additional short-term costs and distant benefits, might outweigh net benefits.

In addition to the reduction in yield, soil degradation can also impose other problems on farmers, generally resulting in an increase in costs at the farm level. These damages include the lower efficiency of fertilizer applications as part of the fertilizer is washed away, higher fertilizer application rates to compensate for the soil fertility loss, elevated stone contents of the fields, an increased need to take new plots into cultivation, etc. (Lutz *et al.* 1994).

To give farmers an estimate of the amount of lost soil that can be tolerated without resulting in a rapid productivity decline a so-called "Tolerable soil loss" has been defined. In Germany, for example, this amount can be calculated by a simple formula, which includes the soil type and ensures that on average not more than 6% of soil fertility is lost in 100 years (Scheffer and Schachtschnabel 1998). In the USA the so-called T-values provide a similar estimate (Erenstein 1999). It gives the farmer an idea of the maximum soil loss tolerable without jeopardizing the ability of the soil to sustain high crop outputs "economically and indefinitely" (Wischmeier and Smith 1978, cited in Pagiola 1994). Nevertheless, due to the site-specificity and the complexity of the erosion problem, in practice these values can provide only a rough estimate. Additionally, they do not help when trying to assess which measures have to be taken and which are the most cost-effective ones to prevent further erosion once soil loss levels have reached a critical point (Pagiola 1994).

While for the individual farmer the amount of soil loss and the costs of soil erosion on his fields might be small, the picture can be different for society as a whole. If erosion is a widespread problem in an area, effects can accumulate. Reasons for this are external effects associated with the off-site effects of erosion. These off-site effects are caused by the deposition of soil after its removal from one location in another site. Clark II (1987, cited in Erenstein 1999) distinguishes between in-stream (in the waterway) and off-stream (before entering or after leaving the waterway) effects. Adverse effects through the transport of soil particles into streams, lakes or the sea and its deposition as sediments can lead to clogging of waterways and reservoirs, affecting also hydro-electric power plants, water storage facilities and navigation. Deposition at coasts can damage coastal habitats like coral reefs. Furthermore, chemical pollutants attached to soil particles, like pesticides or fertilizers, can change the ecology of water bodies, causing for example the eutrophication of lakes.

Off-stream effects include the threat of increased flood occurrence with the subsequent damages to infrastructures and human lives and the deterioration of water quality for human consumption (Wild 1995, Erenstein 1999). All these damaging effects can impose high damage control costs on society, which are nevertheless often difficult to measure or to predict. As soil erosion can be classified as a non-point pollution it is difficult to establish a clear cause-effect relationship and point to one clear source causing the damage (Batie 1986 cited in Erenstein 1999).

On the other hand the deposition of fertile soil in a new location can also have positive external effects. Alluvial plains created through this deposition during periods of flooding are in general especially fertile for crop production after the water has drained. In addition, the bigger sediments can be used as construction materials. The incorporation of these positive externalities in the calculation of the costs of soil erosion for society though is quite difficult and has been often neglected (Erenstein 1999).

2.2 Soil erosion in Central America

Soil erosion and the subsequent loss of soil fertility is a long known problem in Central America[1]. In the following evidence for its extent will be presented, and the main causes of the problem investigated.

1 Due to historic reasons Central America includes Guatemala, Honduras, El Salvador, Nicaragua, Costa Rica and Panama. Belize is usually not included, but sometimes figures are also given for this country. If Mexico is included then the region is called "Mesoamerica".

2.2.1 Evidence for the soil erosion problem in Central America

Regional estimates on the amount of soil erosion in Central America and its effect on soil fertility are scarce. One first hint on the severity of the problem in this region is given by the GLASOD study (ISRIC/UNEP 1991) mentioned earlier. In Figure 2-4 it is shown that about 80% of the agricultural land in Central America is affected by human induced soil degradation. This is the highest percentage worldwide. As the main reason causing the problem, the study named "Agricultural Mismanagement" (Figure 2-2), followed by "Deforestation" and "Overexploitation".

Leonard (1986) collected different estimations on the percentage of land seriously affected by soil erosion or degradation in the Central American countries. He showed some alarming figures (Table 2-2). In 1972, 45% of the land area of El Salvador, one of the smallest and most densely populated countries in the region, was classified as being affected by erosion. Also Guatemala with 25-35% and Costa Rica and Panama with 17% of degraded area showed quite high figures. Even if one argues that these estimates are too high and obtained

Figure 2-4: Percentage of agricultural land affected by human induced soil degradation per geographical region

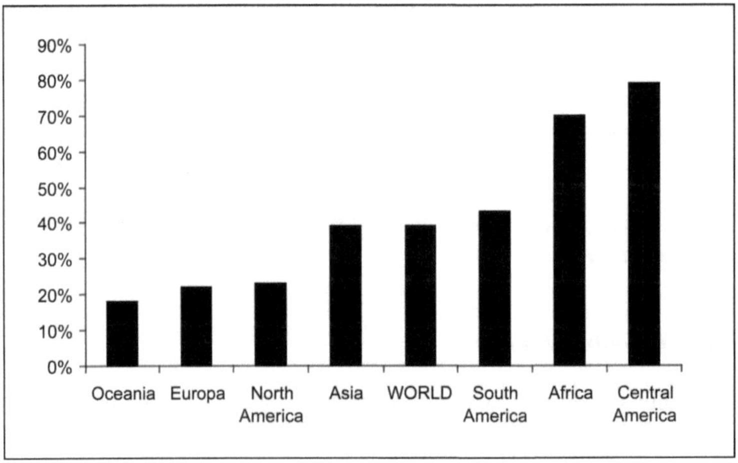

Source: ISRIC/UNEP 1991

Table 2-2: Percentages of land affected by serious soil erosion or degradation in different Central American countries

Country	Percentage	Year of estimate
El Salvador	45	1972
Guatemala	25-35	n.d.*
Costa Rica	17	1981
Panama	17	1980
Honduras	7	1977
Nicaragua	5-10	n.d.
Belize	1	n.d.

* n.d.: no data available
Source: Leonard 1986

with different methodologies, they point in the same direction as the GLASOD Study and demonstrate that soil erosion poses a serious threat to agricultural production and the natural resource base in the region.

Recent empirical evidence on the amount of soil loss is hard to obtain, which is surprising, as soil erosion has raised concerns already for a long time in the region.

Lutz *et al.* (1994) provide in their study on soil conservation in Central America very good insights into different studies carried out on the topic in Central America and the Caribbean (Table 2-3). As the authors stress, these studies cannot be compared easily with each other, because they use different methodologies, and might overestimate the problem by extrapolation from limited data sources and regarding "moved soil" as "lost soil". Nevertheless, most studies show how variable erosion rates can be under different management systems and agro-ecological conditions, and they give a general idea on the amount of soil lost per year. As the case of El Salvador shows the estimated values can be with 137 mT* $ha^{-1}*a^{-1}$ quite high.

Empirical studies on the relationship between soil erosion rates and the decline in soil fertility are also difficult to find for Central America. Estimates are especially difficult, as the calculations often have to be based on parameters from the US with other agro-ecological conditions. Taking data from the US, Wiggins (1981, cited in Ellis-Jones and Sims 1995) estimated for El Salvador a 2% loss of soil productivity for each centimeter of lost soil. As well based on US data, Vásquez (1986 cited in Ellis-Jones and Sims 1995) calculated for Mexico a productivity decline of 15% for a soil loss of two inches (5.08 cm). Lutz *et al.* (1994) give estimates of soil fertility reduction for different locations in Central

Table 2-3: Empirical studies on rates of soil erosion in Central America

Location	Rainfall	Slope	Farming system	Average annual rate of erosion per ha	
	(mm)	(%)		(metric tons)	(mm)
El Salvador, Matapán	1895	-*	Maize	137	8.9
Honduras, Tatumbla 1	2000	45	Maize, beans	42	2.7
Honduras, Tatumbla 2	900-1500	15-40	-	18-30	-
Nicaragua, Cristo Rey	1700	30-40	Cotton	40	-
Panama, Channel area	1200	35	Rice	153	-
Panama, Coclé	1937	-	Rice, maize, yucca, beans	34	17.0
Panama, Chiriquí	1500-2800	-	Pasture	35	5.0

* no data available
Source: adapted from Lutz et al. 1994

America and crops without the use of soil conservation measures (Table 2-4). They calculate for example a yield reduction of 11% after 10 years for coffee plantations in the Heredia region of Costa Rica. After 50 years less than half of the initial yield will be obtained. In the Tatumbla area of Honduras maize yields are estimated to decline much quicker in the first 10 years (by 47%) than in the Yorito region, where the decline is less than half in the same time period (18%). Nevertheless, after about 40 years of continuous maize cropping both sites will arrive at about the same low yields of about 40% of the initial output. For Coco Yam the authors estimate a complete loss in yield in less than 10 years. One can argue though that especially the long-term estimates on yield declines are questionable, as they do not take for example changes in cropping technologies into account. Nevertheless, these exercises are useful to get an idea of the long-term implications of soil erosion for farmers in the region, if no soil protective measures are introduced. Therefore the studies are able to demonstrate the seriousness of the erosion problem in Central America. Further evidence is added by anecdotic accounts from many areas, in which farmers complain about their soil "getting tired".

In summary it can be said that Central America faces a severe soil degradation problem. Evaluating its exact extent though needs further investigation to obtain more detailed evidence on soil loss rates. In addition, it is important to gain further knowledge on its impact on farm output in different environments. Two important questions arise from the evidence found so far: What developments have led to this problem? Thus, why did the GLASOD Study identify "Agricultural Mismanagement", even before "Deforestation", as the main cause

Table 2-4: Estimates of yield reduction for different crops over 50 years for different locations in Central America, without conservation measures (yield in year shown as percentage of initial yield)

Location and Crop	10 years	20 years	30 years	40 years	50 years
Costa Rica, Heredia (Coffee)	89	78	67	56	46
Honduras, Tatumbla (Maize)	53	39	39	39	39
Honduras, Yorito (Maize)	82	65	47	41	41
Costa Rica, Turrubares (Coco Yam)	0	0	0	0	0

Source: Lutz et al. 1994

for the observed problem? And what is and can be done to slow down soil erosion rates, prevent further soil degradation and even reclaim affected soils?

2.2.2 Factors influencing the soil erosion problem in Central America

To understand some of the developments that led to the erosion problem present in all of the Central American countries today one has to look at two main factors:

- The agro-ecological characteristics of the region.
- The land use patterns in the region and the developments leading to the still continuing extension of the agricultural frontier into marginal hillside areas.

2.2.2.1 Agro-ecological characteristics of Central America

As can be seen in Figure 2-5, Central American landscapes are dominated by hillsides and mountainous regions. Table 2-5 shows the percentage of hillside and highland areas for each country. Apart from Belize, all Central American countries have more than 70% of their land area classified as mountainous regions. It can be questioned though if all the percentages given can be compared with each other, as different countries are likely to use different classification criteria. Nevertheless, Figure 2-5 shows that a long band of volcanoes extends from Guatemala down to Panama. Thus some of the soils found in the region are very fertile, especially along the Pacific coast and in the mountainous regions in

Figure 2-5: Map of Central America (darker areas are hillside regions)

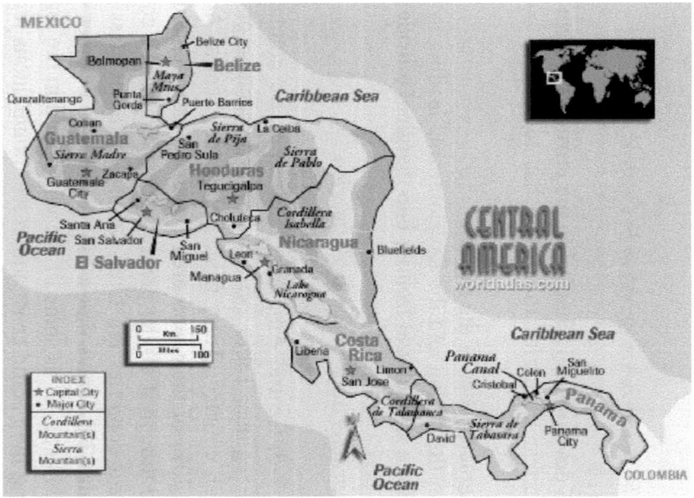

Source: www.worldatlas.com

Table 2-5: Percentage of land area classified as hillside or highland areas

Country	Percentage of land classified as hillsides or highlands
Guatemala	82
Belize	32
Honduras	82
El Salvador	95
Nicaragua	75
Costa Rica	73
Panama	76

Source: Leonard 1986

the interior. But the hilly topography makes their use for agriculture quite difficult and adds to their vulnerability to erosion, once soil cover is removed (Utting 1997).

Central American ecosystems are often characterized by extremes and this also holds true for the soils. Leonard (1986) states that for each hectare of fertile soil about two hectares of poor quality soils with low soil fertility can be found. The poorer soils are located more towards the Caribbean and in northern Guatemala. They are characterized by a thin topsoil layer and are susceptible to nutrient leaching (Utting 1997). Thus, depending on slope length, inclination and soil type land use, especially for agriculture, requires a special management to avoid erosion. In addition, as the breakdown of the organic matter is quite rapid in humid and semi-humid environments, nutrients are mainly stored in the standing vegetation and not in the soil. Thus soil fertility will decline quite rapidly in these soils without a very careful management (de Groot *et al.* 1997).

Also the climate of the region shows a high variability. Rainfall varies between 400 mm $*a^{-1}$ in some drier, subtropical areas towards the Pacific side and up to 7500 mm $*a^{-1}$ in some humid, tropical areas towards the Caribbean (Leonard 1986). The rainy season, which lasts from May to November/ December, is more pronounced on the Pacific side, while more towards the Atlantic side the distinction with the dry season is not so sharp (Utting 1997). Tropical storms and hurricanes often affect the region between July and October (the last example was Hurricane Mitch in 1998, which was estimated to have caused the death of over 10,000 people).

This high variance in topography, soils, and climate results in a large number of different ecosystems and habitats, reaching from humid, tropical rain and cloud forests to subtropical savannahs and dry lands. They provide diverse living spaces for a large number of plant and animal species, which has lead to this region having one of the highest rates of biodiversity in the world.

All the presented factors demonstrate that topographical and ecological factors make agricultural land use in the region difficult, especially the intensive use of hillside areas. Land use systems have to vary as much as the existing agro-ecological conditions and thus have to be adapted to location-specific characteristics. Leonard (1986) estimated that about 25% of the whole Central American land area should be protected from more intensive uses. If this land is taken into production special techniques have to be employed to avoid long-term degradation, if it is possible at all. In general about one third of the mountainous and hillside areas in Central America can be used for annual and perennial crop production. Nevertheless, the area per country varies a lot. El Salvador and Costa Rica are the only countries where intensive annual cropping is suitable for about one fifth of the land area (El Salvador: 24 %, Costa Rica: 19%) (Leonard

1986). In the other countries these areas are much smaller and other land uses, such as the growth of forests or perennial crops, are recommended for the rest of the area. Recommended and actual land use practices nevertheless differ in many cases, as will be investigated in the following in more detail.

Another important issue for understanding the erosion problem in Central America is that, except for some very small parts of Nicaragua, the region does not have natural grasslands suited for livestock production. But as will be shown in the following, pasture for cattle production is the main use of agricultural land in the region. This leads to the conclusion that land is utilized for this purpose, which should have remained forestland or which would be better suited for crop production.

But what are the developments that led to actual land use patterns in Central America today? And how is this related to the erosion problem?

2.2.2.2 Development of land use patterns in Central America and their importance for the soil erosion problem

Figure 2-6 shows land use in the different Central American countries in 1961 and 1994 for forestland and 1961 and 1998 for agricultural land, as a percentage of total land area. Except for Guatemala and Belize, the forest area declined during this time in all countries. In El Salvador, Costa Rica and Nicaragua the drop was especially sharp and reduced the forest area by half. In all of Central America, forest area declined between 1950 and 1990 from 29 million ha to 17 million ha (Kaimowitz 1997). In 1990 a total of 338,000 ha was cleared in the region, which equals to almost 2% of forests and woodlands. Nevertheless, in the 1990s deforestation rates seemed to have slowed down and some agricultural land, mainly pastures, was abandoned and left to the re-growth of woodland and secondary forests (Utting 1997).

Compared with land use suitability, Guatemala and Belize are the only countries that still have land under forest that might also be used for agricultural purposes. In El Salvador, Honduras and Nicaragua land has been cleared which should either be better used as forest or should be utilized only with very special soil management practices. In Costa Rica and Panama the amount of forest area left in 1994 equaled the proposed land use capability (Leonard 1986). Figure 2-7 also shows that forestland was converted into agricultural land, which now in some countries, like El Salvador and Nicaragua, supersedes total forest area. In total about 44% of the Central American land area was used for agriculture in

Figure 2-6: Land use in 1961 and 1994/1998 in Central America (as percentage of land area)

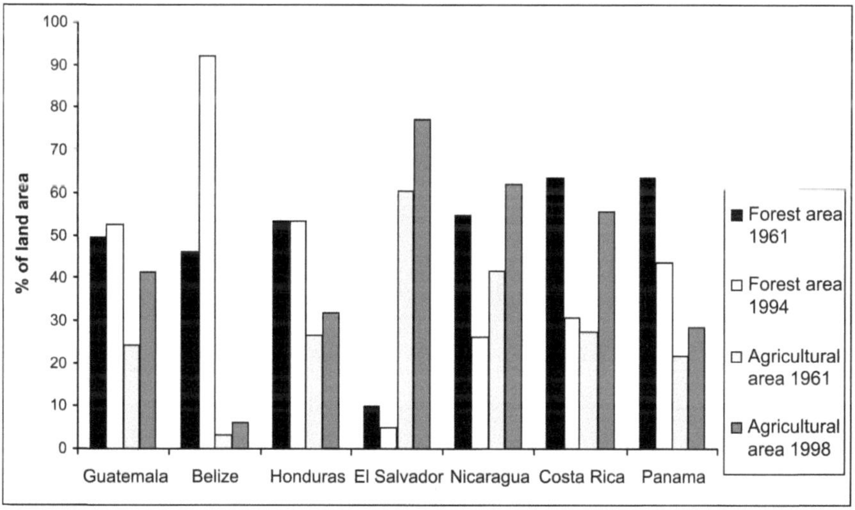

Source: own calculations based on data from FAOSTAT 2000

1998; of this area 31 % was utilized as arable[2] land, 8% for permanent cropping and 61% for pastures (FAOSTAT 2000). Therefore, part of the severe soil erosion problem in the region arises from the conversion of forestland, often in hillside areas, with no or low suitability for crop and livestock production. Another factor that contributes strongly to the problem has been called "Agricultural Mismanagement" in the GLASOD Study. In the following these points are investigated further.

In Figure 2-7 changes in the use of agricultural land can be seen. Except for Honduras and Nicaragua the percentage of arable land has decreased. Some of this land and of the cleared forest area was converted into permanent cropland, which can be seen from the rise in area planted to perennials and plantation crops especially in Belize, Honduras, Nicaragua and El Salvador. From an ecological point of view the change from annual to perennial cropping, which provides a continuous soil cover, can be evaluated as helping to reduce soil erosion. Thus it was found in Costa Rica, for example, that soil erosion can be as much as

2 The FAO (2000) defines "Arable land" as land used for temporary crops, temporary meadows or pastures (less than 5 years), kitchen and market gardens and which is temporarily fallowed.

Figure 2-7: Changes in use of agricultural land in Central America between 1961 and 1998 (arable land, permanent crops and permanent pasture area as percentage of total agricultural area)

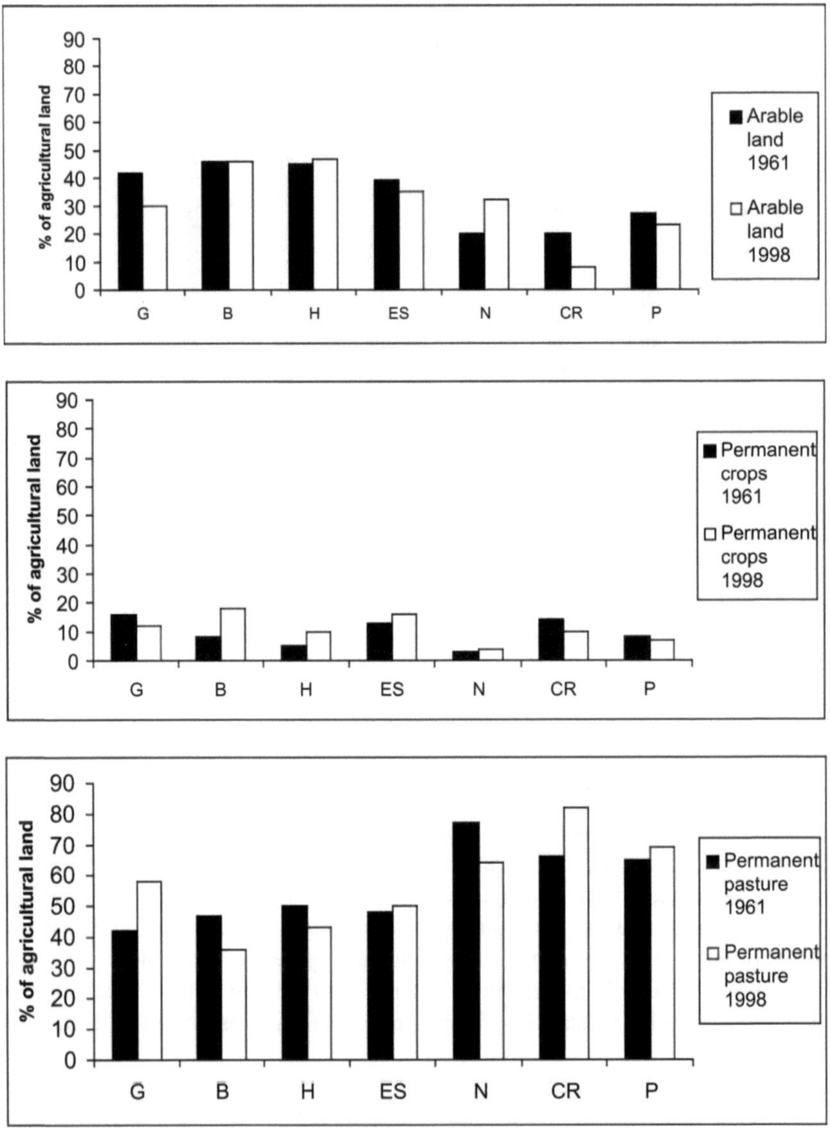

Source: own calculations based on data from FAOSTAT 2000

eight times higher under annual corps than under permanent crops or pasture (Thorpe 1997). The conversion of forestland to plantations though increases the risk of soil loss because until the establishment of a continuous cover the soil is exposed to wind and water. Afterwards the cover is not as complete as a forest cover.

Most of the agricultural land though has been used as permanent pastures[3] for cattle production since the 1950s. Kaimowitz (1997) calls the conversion of forest area to pasture "the most important change in land use in Central America in the last 40 years". This conversion was not a direct one as will be seen later. Land area used for pastures in Central America has risen from 3.5 million ha in 1950 to 6.9 million ha in 1970 and 10.5 million ha in 1983 (Kaimowitz 1996 cited in Comisión centroaméricanade ambiente y desarrollo 1998). In the 1990s the pasture area has declined slightly due to changes in world market prices for beef.

Deforestation in Central America has been associated with the extension of the agricultural frontier into hillsides and mountainous areas and the large rainforest areas of the Caribbean coast. Utting (1997) sees it as a "feature of a particular style of development based on the production of products such as bananas, coffee, cotton and beef destined largely for the international market". This so-called "Agro-export Model" of development was actively promoted by Central American governments already since the end of the 19th century when coffee became the first export product of the region. This in return led to an increasing integration of the Central American economies into the world market (Utting 1997). After the coffee boom followed the expansion of banana, cotton and, in the 1960s and 1970s, cattle production due to the opening to of the US beef market (Howard-Borjas 1995). These developments favored the accumulation of resources in the hands of a relatively small part of society in each country, enforcing existing colonial structures of larger landholdings (*latifundia*) and small-scale peasants (Howard-Borjas 1995). The concentration of crop land in the hands of export producers was one of the consequences. The displacement of small-scale cultivators accompanied this trend, which also resulted in the marginalization of major parts of the population excluding them from decision-making processes and political control (Utting 1997).

A polarized agrarian structure developed comprising a modern, commercialized export sector with a relatively small number of farmers and a subsistence oriented peasant sector producing the main staples, maize and beans. Even today about 63% of the agricultural land in Central America is utilized by 24% of the

3 The FAO (2000) defines "Permanent pasture" land that is used for more than five years for herbaceous forage crops, either cultivated or wild growing.

Table 2-6: The agrarian structure of Central America in 1996

Sector	% of total cultivated land	% of total number of farms
I- Latifundia-Minifundia	(52)	(54)
Extensive cattle ranching	46	10
Subsistence farmers	6	44
II- Agricultural frontier	17	7
III- Modern agricultural sector	17	14
IV- Small commercial producers	14	25

Source: Utting 1996, cited in Comisión centroaméricana de ambiente y desarrollo 1998

farms (Comisión centroaméricana de ambiente y desarrollo 1998). As can be seen in Table 2-6, cattle ranches comprise only 10% of all farms in Central America, while they occupy almost half of the cultivated land. About 44% of the farms, small-scale cultivators, on the other hand only crop 6% of the farmland.

The earlier described developments were also accompanied by a stronger integration of small-scale basic grains producers into the market economy. They now used the main part of their income from surplus production to purchase agricultural inputs and participate in the general trend of higher consumption of non-agricultural goods and services (Utting 1997). An intensification of agricultural production was the result due to the growing need for surplus production and the use of external inputs.

Figure 2-8 depicts the percentage of basic grain producers[4], which are typically small-scale farmers, of the total of all farmers in Central America and of some Central American countries in 1989. Only in Costa Rica these crop producers comprise less than half of all farmers (40%), while in all the countries further north more than 75% belong to this category. Honduras with 90% of the farmers being basic grain producers is the leader in this respect. Thus it becomes clear that with almost 80% of all farmers in Central America being crop producers, while the major part of the agricultural land is utilized by cattle ranchers the distribution of land among farmers is highly skewed and the land use on the remaining land has to be very intensive to produce enough to ensure the survival of the farm family.

Slash-and-burn agriculture, the traditional production system of small-scale farmers in Central America, relies on long fallow periods (6-10 years). This period is needed to restore soil fertility after the land had been used for two

4 In Central America basic grains are maize, beans, sorghum, and rice.

Figure 2-8: Share of basic grains producing farmers in the total of all farmers in Central America in 1989*

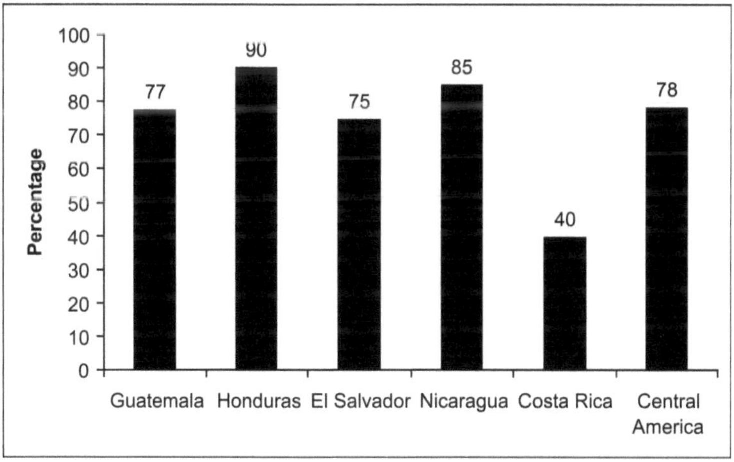

* there were no figures given for Belize and Panama
Source: adapted from Zurek and Sain, forthcoming

to three years. Farmers abandon the fields when fertility declines and clear either a new piece of land or old fallow land. As long as the person to land ratio is not too high, farmers can thus rotate among parcels of land in a larger area and do not have to expand agricultural land deeper into the forest area. This system can be evaluated as sustainable and ecologically sound (deGroot *et al.* 1997) and it can be practiced even in the Central American hillside areas without long-term ecological damage.

Intensification of land use can take place in two forms: by either taking more land into production or by shortening fallow periods. If the latter is the case and fallow periods are shortened or even abandoned completely, so the fields are cropped continuously, the soil productive potential can only be maintained by introducing new soil management practices These include crop rotations, improved fallows, legume intercropping, crop residue management techniques, inorganic and organic fertilizers etc. And especially the burning of crop residues, which makes the soil vulnerable to wind and water erosion and reduces soil organic matter contents, has to be abandoned. In hillside areas the implementation of soil conservation measures becomes a "must", if massive soil erosion and even the land loss due to complete degradation should be avoided. If these measures are not applied, soil resources are "mismanaged", new land has to be taken into production to maintain farm output, resulting in cutting down new forest areas and expanding the agricultural frontier. These developments can be seen in many Central American countries.

The land-intensive slash-and-burn system also breaks down if farmers cannot rotate any more among parcels of land due to increased demand for arable land. This is the case when populations grow and/or when the land in fallow is taken over by other land users as a consequence of insecure land titles. Both phenomena can be observed in Central America.

Accelerated population growth rates (Table 2-7) can be found especially in rural areas of all Central American countries, leading to increased pressure on land resources. In addition, government policies have used colonization programs as a way to alleviate social pressure in urban and older agriculturally used areas, which then raised the person to land ratio at the more remote and agricultural frontier regions (deGroot *et al.* 1997). As governments until recent years perceived forests as an obstacle to development, laws were passed, which granted the right of possession to land only if it was cleared from forest. Nevertheless, formal land titles were often difficult to obtain due to costly, complicated and drawn out procedures, which meant that settlers had to keep the land free from re-growing forest to maintain their claim (Utting 1997, Kaimowitz 1997).

Table 2-7: Population growth rates (percentage per year) in Central America from 1980-1990 and 1990-2000

Country	Growth rate per year in % 1980-1990	Growth rate per year in % 1990-2000
Guatemala	2.8	3.0
Belize	2.8	2.9
Honduras	2.7	3.2
El Salvador	1.2	2.3
Nicaragua	3.1	3.3
Costa Rica	3.3	3.2
Panama	2.3	1.9
Central America	2.6	2.8

Source: own calculations based on data from FAOSTAT 2001

Further factors "pulling" migrants into the new areas were the search for employment in timber extraction or mining activities, the developing infrastructure, and road building (deGroot *et al.* 1997). As better road access facilitates the transport of timber, cattle and dairy products, the provision of a better access is seen by Kaimowitz (1997) as the most important government intervention that added to deforestation and conversion of forest to pasture land. And it increased the attraction of the newly opened areas to migrants. This in return decreased the availability of land for the original population, for whom it became more difficult to rotate among fields. More intensive use of the remaining land was the consequence.

New dwellers were often not acquainted with the agro-ecological conditions in the new area and applied agricultural practices aimed at achieving relatively high outputs with low input levels, as they were lacking capital and faced imperfect input and output markets. Farmers under subsistence conditions tend to have high rates for discounting future benefits received from investments made on their farm. Returns from maintaining or restoring soil fertility though are usually received only after a relatively long time period of several years. This time span together with high discount rates, a low capability of accumulating capital, and insecure land titles lead to a preference of small-scale farmers for low-cost measures, whose benefits can be obtained within a short time period. This in return makes it difficult for them to perceive the benefits of implementing soil conservation practices (this point will be investigated further in the next chapter). Small farmers rather invest any accumulated capital in livestock, especially cattle (deGroot *et al.* 1997). The extractive methods used by many small-scale farmers resulted, not only in the Central American hillside areas (which, as we have seen, comprises most of the land area in the region), in accelerated soil erosion and a rapid decline in the soil productive potential, forcing farmers to clear new land. Or, if no new land is available, to migrate again to a new area or to the cities (deGroot *et al.* 1997).

The second development, which leads to the breakdown of the traditional slash-and-burn system, is the take-over of fallow land by other land users. It can as well be found in many regions and plays an especially important role in the conversion of forests into pastures. DeGroot *et al.* (1997) even call cattle ranching "the logical outcome of agricultural activities in Central and South American rain forest areas". Cattle ranching can be carried out on almost every type of agricultural land, from the very best to the poorest one, and even on relatively steep slopes. This results in a large part of the farmer's income being realized from the land rent (Howard-Borjas 1995), making ranching the most profitable form of land use in many areas. Additionally, ranching requires only a very low labor input (Howard-Borjas 1995) and is characterized in Central America by a relative low technology level (Kaimowitz 1997). Therefore, ranchers, who want to expand their production, expand pasture land.

If land titles are not well defined in an area, one possibility often used is to take over fallow land. As has been described earlier, usually the only way to demonstrate the possession of land is to keep it in production, clear from brushes and forest. But the re-growth of secondary vegetation is necessary for the soil to recuperate, thus crop producers had to leave it to fallow. And usually it is quite difficult for them to enforce their usufruct rights (Howard-Borjas 1995).

Another option for ranchers is to buy land on which crop producers face soil fertility problems after a few seasons, especially as natural pasture species are usually the first to invade fallow land. Clearing of new land or repeated migration are then the options for these farmers. If the crop producers themselves were able to accumulate any capital at all they are likely to invest it in cattle and sow pastures, but not to maintain the soil fertility (deGroot *et al.* 1997).

The result of both described processes is the same: The traditional soil management system with long fallow periods cannot be applied any longer and crop production has to be intensified in the remaining areas. Crop production is even pushed into marginal lands, which are the hillside areas, and more and more land has to be taken into production. Applied practices though are usually not adequate, especially for hillside regions, as slash-and-burn methods continue. Soils are mismanaged and soil fertility declines. This in return leads to further deforestation and expansion of the agricultural frontier, if possible. As the most visible consequence severe soil erosion and land degradation can be observed.

Going hand in hand with a worldwide discussion about the preservation of natural capital, the 1990s brought a decline in deforestation rates in Central America and a general change in thinking about environmental issues (Utting 1997, Kaimowitz 1997). Utting (1997) suggests that two developments are responsible for this: First, a growing concern of governmental, non-governmental, and development organizations about worsening environmental degradation in the region. This resulted not only in an increase in environmentally oriented projects and aid, but also in changes of legislation and policies to protect forest resources, encourage reforestation and slow the expansion of the agricultural frontier. In addition, the protection of biodiversity and genetic resources, which are of interest especially for international pharmaceutical companies, has become an important issue in recent years. Second, the business sector started to take a rising interest in the provision of environmental goods and services, profiting from governmental incentives for reforestation and forest plantations, sustainable logging, environmental planning and 'eco-tourism' (Utting 1997). Recently also the provision of carbon fixation schemes for developed countries to offset their carbon emissions has become a new field for business opportunities.

All these developments - badly needed to preserve the remaining forests in Central America and raise the awareness about the environmental degradation in the region - also highlight the necessity for intensifying land use on the remaining agricultural land, as new land should not be taken into production. Accelerated soil erosion rates have been an issue of concern for the last 40 years (see for example, Lutz *et al.* 1994, Cuesta 1994 or Hernández Navas *et al.* 1994) and all Central American governments encouraged the implementation of a large number of national and international projects and programs to develop and promote new soil conserving management practices. This was also related to the issue of how continuing soil degradation might threaten the survival of small-scale, subsistence producers in the region . In the following a short overview of promoted soil conservation techniques will be given to describe the available options.

2.2.3 Towards a classification of soil conservation practices promoted in Central America

The number of projects and programs working in the field of developing and promoting new soil management practices in Central America seems to be quite high. In 1996 Dvorak, working for the CIAT Hillside Program in Honduras, compiled a comprehensive list of soil conservation projects carried out in Central America and Mexico (Dvorak 1996). Most of the following deliberations are based on this list.

In total Dvorak (1996) found 91 projects working with different soil conserving practices. Figure 2-9 depicts the percentage of investigated projects working in the different Central American countries and Mexico. As can be seen, conservation projects can be found in all investigated countries. The majority of them are located in Honduras (37%), while about one fifth of them work in Nicaragua followed by Costa Rica (11%) and El Salvador (10%). In the other Central American countries and Mexico the number of conservation projects is less than 8%.

Table 2-8 gives an overview of different soil conservation practices promoted by the projects, which Dvorak investigated. In addition, the share of each practice in the total of all promoted techniques is presented. These practices can be categorized in different ways: in terms of how they change soil management practices (this is the categorization which will be used for this study), in terms of how they operate to reduce soil loss, and according to certain economic characteristics.

Figure 2-9: Distribution of soil conservation projects from Dvorak's list among Central American countries and Mexico

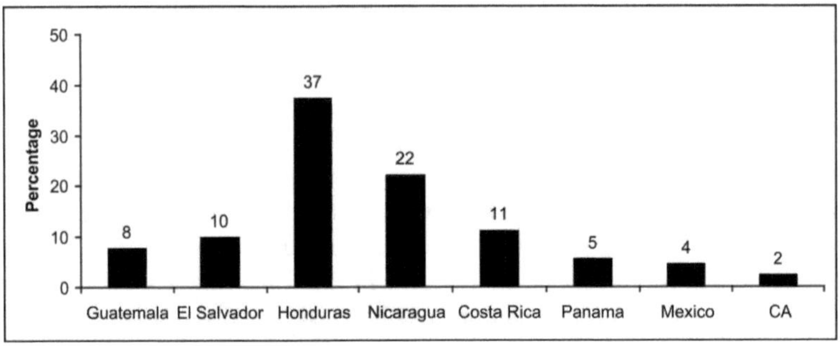

Source: adapted from Zurek and Sain, forthcoming

If soil conservation techniques are classified according to the changes they imply for soil management, these practices can be divided into three categories:

- Physical structures: Individual and bench terraces, dikes, drainage, drainage ditches, contour ditches, contour ridges, dead barriers.
- Techniques that imply a direct change in soil management practices: Ridge tillage, live barriers, no-burning, minimum tillage, crop residue management, green manures/cover crops, improved fallow, improved pasture, windbreaks, living fences.
- Complex land use systems: Silviculture, agroforestry (for this: tree and communal nurseries).

In Table 2-8 it can be seen that the most popular techniques among the investigated projects belong to the category of changes in soil management practices, namely no-burning (11.4%), live barriers (11.0%), green manures/cover crops (10.4%) and crop residue management systems (8.1%). Physical structures are not promoted as widely. In this category dead barriers (7.2%), drainage channels (6.8%), and individual terraces (5.5%) are the most prevalent ones. Complex land use systems only play a marginal role; here only reforestation schemes as part of silviculture are of some importance. All the other offered soil conservation techniques are of minor importance, though they are probably often promoted together with the more popular ones.

These practices can equally be categorized according to the way they work to prevent or cure soil erosion (Figure 2-10). Preventive measures, which try to avoid the movement of soil, either reduce the impact of rain or run-off by providing a thorough cover for the soil. Most practices in this group belong to the category of changes in soil management, like no-burning of crop residues, green

Table 2-8: Soil conservation practices promoted by projects in Central America and Mexico according to the Dvorak list

Category of soil conservation techniques	Soil conservation technique	% of investigated conservation projects promoting technique
Physical Structures	Individual terraces	5.5
	Bench terraces	2.1
	Dikes	2.5
	Drainage systems	1.3
	Drainage channels	6.8
	Drainage channels along contours	0.8
	Contour ditches	3.4
	Contour ridges	0.4
	Dead barriers	7.2
Changes in soil management practices	Ridge tillage	5.5
	Live barriers	11.0
	No-burning practices	11.4
	Minimum tillage	2.1
	Crop residue management	8.1
	Green manure/Cover crops	10.2
	Improved fallow	0.4
	Improved pasture	1.3
Complex land use systems	Silviculture	5.9
	Agroforestry	0.8
Other related practices	Tree nurseries	1.3
	Communal nurseries	3.0
	Windbreaks	0.8
	Living fences	2.5

Source: adapted from Dvorak 1996

manures/cover crops or improved fallow. But also complex soil management systems, like agroforestry and silviculture, have preventive qualities. The second group of preventive practices tries to lower the erodibility of the soil by maintaining its aggregate structure and moving it as little as possible. This group again is comprised of some changed soil management techniques, like minimum and zero tillage practices[5].

5 In Central America the typical soil management practice of small-scale farmers is a zero tillage system, in which the farmer does not prepare a seedbed at all, but only makes a hole with a planting stick into which the seeds are placed.

Figure 2-10: Classification of soil conservation practices promoted by projects in Central America and Mexico according to their preventive or corrective qualities

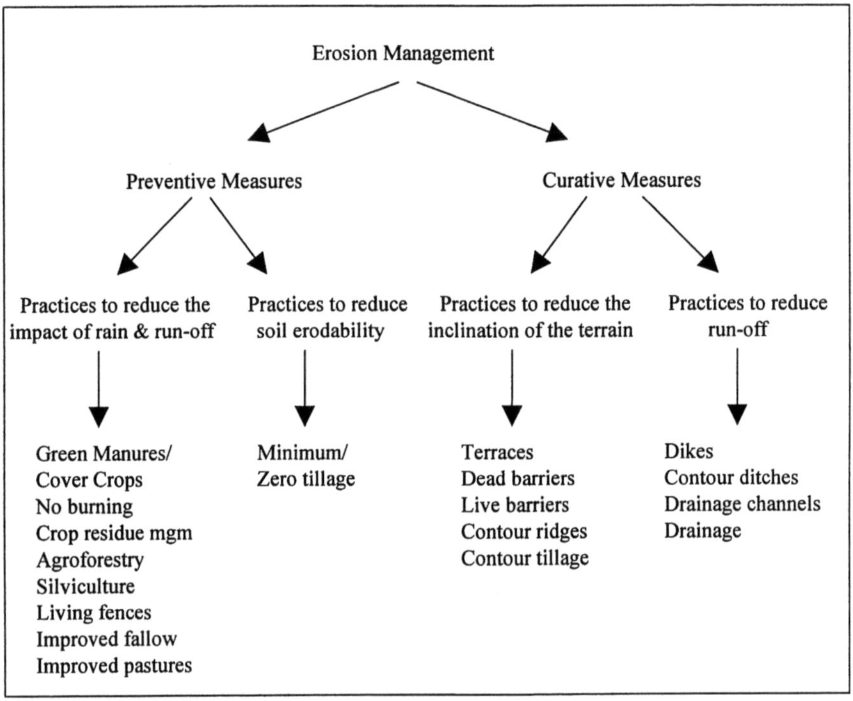

Source: adapted from Lal 1995

In contrast curative measures try to control the damage once erosion has become a problem and to manage run-off. Except for live barriers, all techniques included here belong to the category of physical structures. Corrective measures either reduce the inclination of the terrain by using terraces, live or dead barriers that retain the lost soil, and contour ridges and tillage. Alternatively they reduce the amount and velocity of run-off and channel the water by implementing drainage systems and ditches and channels along and perpendicular to the slope (Lal 1995).

The described soil conservation practices differ greatly in their complexity, their input use and their maintenance costs, when analyzed from an economic perspective. Therefore another way to group soil conservation techniques might be according to their economic characteristics. Zurek and Sain (forthcoming) attempted an economic classification of 16 soil conservation practices promoted in Central America. For this classification the authors used the following classification criteria: Cost-benefit profile, impact on risk, intensity of production factor use, and complexity.

Results suggest that most physical structures, like terraces, dikes or drainage channels, require a substantial initial investment in labor and/or capital for material and hired labor. Therefore they can be classified as capital and labor intensive. In addition, it usually takes more than two vegetation cycles until benefits are received because investment costs have to be recuperated and yields increase only slowly, if no additional measures are introduced that restore soil properties. The implementation and correct maintenance of many physical structures also requires technical skills and knowledge and can be quite complex.

Changes in soil management practices do not require large initial investments, though they demand a basic understanding of agro-ecological relationships within the production system. Nevertheless, they are in general less complex and easier to maintain than physical structures. Especially in the case of practices that provide additional organic matter to the soil (cover crops or crop residue management techniques), the time period is quite short until first benefits through yield increases are received by the farmer. And often they help to decrease the risk of crop failure in case of a bad year. These favorable characteristics provide an important incentive for adoption. Nevertheless, when compared with the traditional slash-and-burn farming system prevalent in Central America, some of the changes in soil management practices can result in higher labor or land requirements. Therefore, depending on farmers' circumstances these different input requirements can result in an obstacle to adoption of the new practices; a point that will be investigated in more detail in the following chapters.

Complex soil management systems, such as agroforesty or silviculture, require on the one hand large investments in knowledge acquisition, as they are often quite different from the traditional farming systems. Farmers also have to invest capital to obtain the necessary inputs such as specific tree species. The availability of these inputs can often be a major obstacle to adoption. In addition, the complexity of these new systems makes it hard for the farmer to adopt only parts of them to minimize the implementation risk at the beginning.

All the characteristics of the described soil conservation practices heavily influence their profitability when implemented and the choices farmers face when deciding which practices to use. All measures are quite effective in reducing the amount of lost soil within a few cropping periods, but they also have a number of other characteristics that do not always suit farmers' interest. And seeing an impact on farm productivity through an increase in crop output can take more than a few cropping seasons. However, the main aim of these practices is to protect natural resources and improve the way they are managed. Therefore, they have sometimes been called 'Environmental Innovations' (Lawrence and Vanclay 1994). Of course they also aim to improve farm output in the long-run, but the main focus lies on preserving natural capital. This is one of the points in which these techniques differ from so-called 'Commercial Innovations', such as fertilizer or tractors, whose effects on crop productivity can be seen usually within one cropping cycle. One of the main hypothesis examined in this study is that the differences between these two types of technologies can have important consequences for farmers' adoption behavior.

In the following chapters the differences between environmental and commercial innovations and the consequences for farmers adoption behavior will be investigated in detail. It should be noted here that there is a wide range of options that can be used to mitigate soil erosion and each practice can be classified according to different characteristics. Many organizations and projects also promote a combination of different techniques (e.g. no-burning practices with live barriers and cover crops). Figure 2-11 shows that about two thirds of the investigated projects work with more than one practice.

Understanding the different solutions to the soil erosion problem nevertheless confronts us with the question of who are the people to whom conservation practices are offered and who are supposed to adopt these techniques.

Figure 2-11: Number of soil conservation practices promoted by projects from Dvorak list

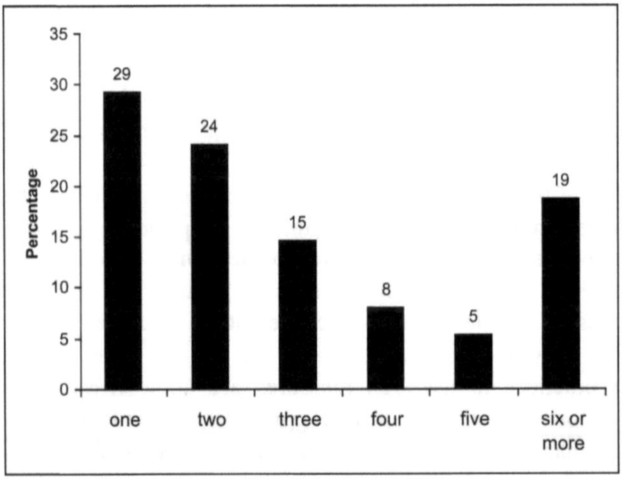

Source: Zurek and Sain, forthcoming

As described earlier, the main crop producers in Central America are small-scale, subsistence farmers, who are using slash-and-burn practices to produce their most important staples, maize and beans, often in marginal and hillside areas. These are the farmers, who cultivate the ecologically most fragile lands in the region. Unfortunately they often use inappropriate methods, and hence cause a substantial part of the erosion problem. This then results in these crop producers being the main target population for the promotion of soil conservation techniques. But despite all the efforts described earlier, undertaken already for several decades, the overall adoption of these practices is relatively low (Lutz *et al.* 1994, Thorpe 1997). Thorpe (1997) cites the last Honduran census, in which only 7.1% of the farmers with 5-10 ha of land and 10.1% with more than 10 ha implemented soil conservation practices. For other countries these figures are probably similar, though it is difficult to obtain any accurate national estimates. Nevertheless, there are also some successful local examples for the adoption of conservation techniques: One is the case of the adoption of no-burning practices together with crop residue management (here called conservation tillage), live and dead barriers in a package with improved maize and sorghum varieties and fertilizer in Guaymango, El Salvador. Adoption rates have been reported to reach almost 100% in 1986 (Sain and Barreto 1996). Another example is the use of the green manure/cover crop Mucuna in the Litoral Atlantico region of Honduras, where adoption reached 64% in 1992 (Buckles *et al.* 1998). But unfortunately, these examples cannot be found often in the region.

Many projects and programs tried to foster the adoption of soil conservation practices by providing direct incentives for adoption. These included food for work programs, payments, the supply of inputs like seed material, fertilizer, tools etc. or credit programs. Of the 91 projects investigated by Dvorak (1996) only 24% did not use any incentives of this kind (Zurek and Sain, forthcoming). And also Rosado *et al.* (1994) and Hernandez Navas *et al.* (1994), to give only a few examples, reported the use of these measures for Guatemalan and El Salvadorian conservation projects, respectively. But they also reported that there exist serious doubts about the usefulness of these incentives, as the implemented practices were often abandoned as soon as the incentives were withdrawn. Giger (1999) provides a comprehensive report about the problems associated with these measures in order to foster the sustainable adoption of conservation technologies. He concluded that direct incentives can even be counterproductive.

In summary it can be said that quite a wide range of soil conservation techniques exist in Central America, which possess different agronomic characteristics concerning the reduction of erosion and all result in a measurable decrease of soil loss. Their use is promoted mainly among small-scale farmers, the main crop producers in the region. But despite many long-lasting efforts and direct incentives provided to farmers there are few examples of successful long-term use of these practices and their overall adoption levels are relatively low.

It becomes clear that some other factors exist, apart from the proven technical effectiveness of the techniques, which influence the adoption behavior of small-scale farmers in Central America. In this study some of the reasons for these low adoption rates of soil conservation techniques will be examined. The "Induced Innovation Theory of Technical and Institutional Change" is utilized as the theoretical background for investigating some of the principles underlying technology development and adoption. This theory was proposed by Yujiro Hayami and Vernon Ruttan in the 1980s as the first consistent model to include technical and institutional progress in a society as endogenous variables of the development process. In this context also the economic factors influencing the adoption of a successful conservation practice by small-scale farmers have to be investigated in order to better understand farmers demand with respect to new technologies and draw conclusions for technology development and policy design (Chapter 4). In addition, the way in which conservation practices are selected and promoted among the target population by organizations working in this field is studied in order to investigate the supply side of conservation practices in Central America (Chapter 5).

3 Environmental Innovations and Technical Change

The low adoption of soil conserving practices in Central America is a long-lasting and pressing concern of conservationists in the region, as it not only threatens the regions natural resource base but also the survival of many small-scale farmers. The fact that many of the efforts employed to raise adoption levels have shown only limited results, raises the question of how technical change in new environmental innovations can take place in a socio-economic and institutional environment as it can be found in Central American countries. In particular: What are the important factors influencing the decision making process of farmers concerning these new technologies? What characteristics different from other agricultural innovations do these practices possess? And what are the policy and institutional implications concerning the system for generating new environmentally sound practices in Central America?

A first step that can help to answer these questions is to categorize new technical innovations as either Commercial Innovations (CIs) or Environmental Innovations (EIs) and to lay out the implications of some of the characteristics of EIs for farmers' adoption behavior. In a second step, the theoretical background of technical change and the adoption of new technologies in the agricultural sector will be investigated to see if the described mechanisms for inducing the implementation of CI technologies can be applied to EIs. For this question one of the most comprehensive theories of technical and institutional change, the "Induced Innovation Theory" proposed by Hayami and Ruttan, was chosen. In this context the processes described in the Theory for explaining the interactions between technology suppliers and users are studied with respect to EIs.

3.1 Differences between Environmental and Commercial Innovations

Commercial (CI) and environmental innovations (EI) can be differentiated in a few decisive points that have important consequences for their implementation by farmers.

CIs are developed with the purpose of increasing farm productivity and with this farmers' income. They are thought to be usable by all farmers under similar circumstances (Vanclay and Lawrence 1994). High-yielding varieties and fertilizer can serve as examples of technologies that raise land productivity, while tractors augment the area a worker is able to farm (Hayami and Ruttan 1985). In this manner, either output is increased or input costs are reduced, resulting in overall higher net income for the farm operator. Thus the adoption of these new technologies is seen as raising farmer's utility. The classical model to understand the adoption process is the so-called diffusion model, developed by Rogers and others. It depicts the relationship between adoption level and time after promotion of the new practices in a non-linear cumulative growth curve. Depending on the farmer's resource endowments, skills and his inclination to experiment farmers are described either as innovators, early adopters, early majority, majority, late majority or laggards (Rogers 1983). The implication of the model is that if the innovation provides proven economic benefits to farmers, usually in a relatively short period, and if farmers behave according to a profit-maximizing rationale, adoption will occur. If adoption does not occur it is only the farmer himself who is affected as he forgoes profit. Non-adoption of the new, benefit increasing practice is usually thought to occur due to risk-averse behavior of farmers that first have to be certain of possible gains or because the techniques do not fit farmers' circumstances, including his resource endowment, socio-economic or cultural background or his knowledge or educational skills. In other words, for some farmers non-adoption can make economic sense (Vanclay and Lawrence 1994).

EIs in contrast are defined by Vanclay and Lawrence (1994, p. 3) as ".... the use of techniques, methods and approaches to improve land management rather than to increase farm productivity...". Innovations such as soil conservation techniques, integrated pest management, etc., are designed primarily to protect the natural resources -soil, water, genetic resources, etc.- serving as the basis of the agricultural production process. The vision behind these practices is a long-term one, encompassing the idea of recursive interactions between soil, water, plants and livestock, which in return influence long-term performance of the agricultural system. The implementation of EIs might thus imply that the farmer has to forgo short-term economic benefits in order to reach a healthy agro-ecosystem for his own benefit in the longer future and for the benefit of society as a whole that is interested in a healthy environment. In their critique on the use of the traditional extension approach for the promotion of environmentally sound practices, Vanclay and Lawrence (1994) include a comprehensive description of characteristics of EIs. The following description of EI characteristics is partly based on their deliberations and is complemented by other sources.

- EIs are usually not just new additions to the already known farm management practices but encompass a whole new system of thinking about the interactions between soil, water, plants and livestock under different natural conditions. Therefore their implementation often implies profound changes in the whole management system and in the need to understand the complex linkages within the system. Farmers have to be aware of their own impact on changes among these linkages to use them appropriately without destroying the interactions needed to sustain agricultural production.

- Adjusting farm management practices in order to sustain the long-term functioning of the complex interactions in the agro-ecological system implies that these adjustments can differ depending on the kind and strength of relationships present. This in return signifies that not all new environmentally sound practices are equally applicable under all conditions, resulting in the need to devise location specific adaptations in practice. Thus Vanclay and Lawrence (1994) suggest that the adoption of a particular technology does not necessarily give an indication of the degree of "environmental soundness" reached by its implementation. It might be better to consider the degree of environmental protection achieved rather than the adoption of an environmentally sound technology per se.
- Implementing a more systemic approach for farm management can have two additional implications: First, often EIs can be relatively complex, making them difficult to understand and implement. The farmer is required to possess certain knowledge of the underlying processes within the system and often he needs to have some particular management skills. But this increases the difficulties and the risk associated with the implementation of the new technology. This in return can result in farmers' resistance to adoption of the practices. Second, the more complex the new environmentally sound technology is, the less likely it is that it can be divided into smaller steps, which the farmer can first try before implementing the whole new set of practices. Thus partial adoption as a way to minimize the adoption risk is less likely to occur, which in return can impede the adoption of the technology all-together.
- One fundamental problem of many EIs is that establishing the new farm management practices can be quite labor and/or capital intensive. In addition, yields often decline for an extended period of time after the implementation of the new technologies. Yields will start rising again once the damaged agro-ecological interactions of the old farming system are recovered. This can result in a considerable time lag between the occurrence of costs of the new technologies and receiving benefits. Costs can even completely outweigh benefits on the individual level, depending on which kind of technology is used and the time horizon utilized to assess overall costs and benefits (Pagiola 1994). In this case a farmer acting according to profit-maximizing principles is not very likely to adopt the new practices. This adds evidence to the existence of diverging social and private returns to the use of

environmentally sound practices (Lutz *et al.* 1994). In the preceding chapter the emergence of off-site damages from enhanced soil erosion was described. This can lead to costs imposed on society as a whole when forced to take measures to mitigate these damages. Therefore, from a societal point of view it is desirable that farmers introduce environmentally sound practices, such as soil conservation technologies, to reduce the detrimental external effects of the other practices. The costs of implementing and maintaining these practices though, have to be borne by the individual farmer, thus reducing farm net benefit, at least at the beginning (Lutz *et al.* 1994). Society, or governments, might then have to think about compensation or other measures to induce the implementation of environmentally sound practices. Technology suppliers might have to reconsider the new techniques offered to farmers in order to fully understand different technology characteristics and related farmers' adoption behavior. This then could help with tailoring these practices better to farmers' circumstances .

- A large number of EIs developed so far fall in the category of agronomic management practices, which often use locally available materials. There are fewer technical/mechanical or biological innovations. Thus most EIs consist in new information that is provided to the farmers, while fewer are directly embodied in equipment, which can be sold to farmers. The public good character of information and thus of many EIs leads to few private technology suppliers being interested in their development. Therefore, so far most of the resources spent by private companies for agricultural research are allocated to crop improvement research while very little is invested in the development of new natural resource management techniques (Pingali, in review). If farmers are not able or willing to pay for the provided information it will be difficult for them to recuperate their research costs. Here public research and development (R&D) organizations have an important role to play. Additionally, the use of EIs is often not just in farmers' interest but also in society's. This raises the question why farmers should pay not only for implementation costs but additionally for the development of EIs. Here again it can be justified why the research carried out on this kind of technologies should be publicly funded.
- In addition, to achieve a certain degree of environmental protection it is not enough to consider the behavior of only one farmer. The ecological system employed by one farmer does not stop with farm borders and often management practices of one farmer can have an impact on ecological interactions in the system of a neighboring farm. Reaching a reduction of environmental degradation can imply the need for collaborative efforts of many farmers in an area, e.g. a watershed. This requires not only a certain degree of self-

organization and communication among farmers, but also all involved farmers must be aware of the influence of their actions on the community and must feel the need for change. Here the disposition of an individual farmer to consider trade-offs between personal benefits associated with the use of degradation causing management practices and costs imposed on him when changing these practices for his AND societies benefit becomes important. Environmental protection and the intrinsic value assigned by a society to the existence of well functioning ecosystems thus turn out to be an issue. For a farmer to really act also on behalf of society the use of environmentally sound practices has to be socially accepted in an area as good farm management. Additionally, the motivation of farmers to act in the interest of society and the incentive mechanism that can induce them to forego short-term profits in order to contribute to the well-being of all need to be considered.

- Assuming that farmers are willing to lose a certain amount of their direct, short-term benefits in return for receiving benefits from a well-functioning agro-ecosystem in the future and for contributing to society's benefit the question arises of how large of a loss can be absorbed by the farmer. Under subsistence conditions this amount is likely to be almost zero. This leads us to the question if EIs exist or can be developed that combine environmental protection with increasing farm profitability in both the short and the long-run. Thus the incentive for farmers to use these practices would not only arise from their willingness to contribute to the reduction of environmental degradation but it would also fit into the economic rationale of profit-maximizing behavior. Furthermore, the question arises of how technical changes in agricultural practices that are congruent with environmental concepts can be understood in the context of the "Induced Innovation Theory". These two issues will be investigated in further detail in the following.
- The likelihood that adoption of environmentally sound practices occurs is not influenced only by perceived costs and expected benefits. The farmer might also find that certain characteristics of the new practices do not go along with his resource endowments, knowledge and educational skills or cultural beliefs. Taking the example of soil conservation practices, a farmer might find that the implementation of terraces requires a lot of labor and engineering skill not available to him. In addition, the implementation of conservation practices can reduce the land area that can be cultivated or draw on the farmer's capital resources, which he might require to cover certain consumption needs of his family. Here it becomes clear that even if a farmer wants to introduce environmentally sound practices on the farm he might not be able to do so due to incompatibilities with the existing infrastructure and/or the conditions in which farmer takes his adoption decisions.

- All the described differences between CIs and EIs result in the need to employ different methods and policy instruments to foster the adoption of either of the two technology groups. For CIs it might only be necessary to adjust outer farm circumstances (e.g. by providing credit) to help facilitate widespread, voluntary adoption. As we will see later, according to the "Induced Innovation Theory" though, this kind of state intervention is not even needed if markets work properly and technologies are tailored to farmers' conditions. These measures can even impede long-term growth if input and output prices are distorted too much, thus giving wrong signals to farmers and technology developers. For EIs on the other hand measures like subsidies or laws and state regulations have often been the first choice. Many decision-makers have seen subsidies as a compensation for costs arising from the implementation of EIs, which in return should induce farmers to voluntarily adopt. Laws and state regulations on the other side obligate farmers to use EIs. The need to use either one of these approaches shows that most currently available EIs do not seem too attractive to farmers, something that was already mentioned when describing the low adoption of soil conservation practices in Central America in the preceding chapter.

From the described differences between EIs and CIs it becomes clear that the adoption of EIs is not as much influenced by short-term profit-maximizing considerations as might be the case with CIs. The main motivator for the development and application of these technologies is considered to be reaching a reduction of environmental degradation, which will help to increase farm output and farmer's utility in the longer run on the one side, and contribute even more to society's well-being by providing a healthy environment on the other side. Thus the traditional ideas on the adoption of new practices and on analyzing the factors influencing adoption might have to be reconsidered in this context.

But how can we reconsider farmers' adoption behavior with respect to EIs? What can we learn from the models developed to explain technical change in the agricultural sector in understanding this question? Is it possible to incorporate the farmer's perception on how much environmental degradation might influence farm productivity and with this the survival of his family in the long run into these models? Can these thoughts be coupled with an understanding of the importance of the farmer's behavior for his community or society as a whole? The question is if these two motivators are enough to induce a switch to new farm management practices that are also associated with various difficulties and with the loss of short-term profits. Furthermore, the farmer has to have a long-term perspective, which is often difficult under subsistence conditions of small-scale farmers in developing countries, and he has to be aware of the interactions between degradation and soil fertility/overall farm output. Additionally the question arises if the technology development processes laid out so far by the models on technical change can help us to understand the interactions between technology suppliers and users necessary to develop the kind of EIs that can be implemented easily by farmers.

3.2 Induced Innovation Theory of Technical and Institutional Change and Environmental Innovations

Different theories exist about how technical change and the creation and adoption of innovations occur in firms. These include the Evolutionary Theory, which is based on the notion that firms behave according to existing decision rules when searching for new technologies, and Path Dependent Models of innovation. In this theory the choice of new technological options depends on the offer and the decision taken in the past (Hussain unpublished).

One of the first and most widely accepted models describing a consistent framework of technical and institutional change in the agricultural sector and illustrating the process of agricultural development, is the so-called "Induced Innovation Theory of Technical and Institutional Change" put forward in 1971 and, in an extended form, in 1985 by Yujiro Hayami and Vernon W. Ruttan (Hayami and Ruttan 1971, Hayami and Ruttan 1985). The following explanations of the theory mainly draw on the extended version.

The authors define technical change "...as any change in the production coefficients resulting from the purposeful resource-using activity directed to the development of new knowledge embodied in the designs, materials, or organizations" (Hayami and Ruttan 1985, p. 86). The theory incorporates elements of other theories such as: J. R. Hick's Theory of Wages (Hicks 1932), stating how changes in relative prices of production factors can influence the direction of technical change; ideas of Z. Grilliches (Griliches 1957) and J. Schmookler (Schmookler 1966) on changes in the rate of technical change due to growth in product demand; S. Ahmad's (Ahmad 1966) and H. Binswanger´s (Binswanger 1974) research production function model and innovation possibility curve, which encompasses the technologies that can potentially be developed with a certain research budget at a given point in time; and T. W. Schultz' (Schultz 1964) High-Payoff Input Model, which describes the importance of supplying high payoff inputs to farmers and which stresses the need for investments in human capital. Different from other theories on economic growth, that treated the development of new technologies as a process independent of other advancement processes in a national economy, the proposed model is able to introduce technical and institutional change as variables endogenous to the development process. It recognizes the existence of differing paths of development under different resource, institutional and cultural endowments, leading to different technical solutions to release farmers' resource constraints. Here farmers' search for possibilities to substitute a relatively abundant production factor for a relatively scarce one is seen as the fundamental mechanism, inducing in return public and private technology suppliers to provide a specific set of technological options (Hayami and Ruttan 1985). Pre-requisite for this is the smooth functioning of input and output markets to provide appropriate price signals, as well as an effective research and development (R&D) system capable of responding to signals coming from technology users.

The "Induced Innovation Theory" combines different mechanisms working at various levels to explain the occurrence and the direction of technical change. On the one side the behavior of the individual farmer with respect to the adoption of new technologies is explained, using certain assumptions on farmers' decision-making rationale. The farmer's behavior is then linked to the wider macroeconomic and institutional context, which leads to the description on how technical change takes a certain direction in economies with varying resource endowments. Here then the mechanisms for developing new technical options that are usable under the prevailing resource constraints in an area and the investigation of the interactions between technology supplier and users is incorporated into the Theory.

Most research carried out on testing the theory and its applicability under different institutional and cultural settings focused on the development and adoption of CIs like high-yielding varieties, fertilizer, irrigation systems, tractors and other machinery, etc. As described earlier all these technologies are aimed at either increasing agricultural output in a relatively short period of time (usually already in the period of their implementation) or reducing total input costs. Thus farmers' adoption behavior with respect to these new practices is expected to be governed by a general profit-maximizing rationale consistent with neo-classical microeconomic theory. In contrast to this, many EIs are designed primarily to protect the natural resources -soil, water, genetic diversity, etc.- serving as the basis of the agricultural production process with farm level profitability assuming a role of secondary importance in technology design. Implementation of EIs can imply that the farmer has to forgo short-term economic benefits in order to reach a healthy agro-ecosystem for his own benefit in the longer future and for the benefit of society as a whole.

So how does the development of these techniques fit into the "Induced Innovation Theory" framework? Can the mechanisms governing the adoption of new practices and the interactions between farmers and the R&D system of a country be applied to EIs as well? Or are there other mechanisms needed to facilitate the communication between farmers and technology developers for designing successful EIs?

In the following we will look at the main mechanisms of the "Induced Innovation Theory" and investigate how they apply to the development and adoption of EIs.

3.2.1 The inducement mechanism for technical change in the "Induced Innovation Theory"

As a first step to investigate the development and adoption of EIs in the context of the "Induced Innovation Theory" we will look at the mechanisms laid out to explain the behavior of individual farmers with respect to the adoption of new technologies.

In their explanation of the model of induced technical change Hayami and Ruttan (1985, p. 90) state that "the classical problem of resource allocation...... is, in this context, treated as central to the agricultural development process." During the production process the farmer has to take two interlinked decisions: The first one is on how to allocate the resources available to him in form of the production factors land, labor and capital in the best possible manner to achieve the maximum output. The possible relationships between the production factors depend on the technologies employed by the farmer, as each set of practices requires a certain combination of land, labor and capital. The second decision is on how to achieve this output at the lowest possible cost level. A production factor that is relatively scarce to the farmer will have a higher price for him than a factor that is relatively more abundant. Thus the farmer will try to save the more expensive factor in relation to the cheaper one in order to keep total cost at the lowest possible level. Thus a farmer is likely to adopt a new technology if the new set of practices is able to reduce input cost while providing the same level of output as the old technology, thus increasing farmers' utility.

Changes in relative prices of production factors can lead to biases of technology users in looking for new technology options. In a dynamic economic system these changes are the norm rather than the exception. Drawing on J. R. Hicks' Theory of Wages, Hayami and Ruttan (1985) describe how changes in the demand for certain agricultural products can result in new production factor demand curves. As production factors though have different supply elasticities, depending on farmers' resource endowment, changes in factor demand will lead to changes in relative prices of these inputs. Farmers will try to substitute a more abundant and thus cheaper factor for a scarcer one by switching to new practices.

Thus in the "Induced Innovation Theory" "...technical change represents a dynamic response to changes in resource endowments and to growth in demand" (Hayami and Ruttan 1985, p.84). Hayami and Ruttan (1985, p. 84) want "...to interpret the process of technical change as endogenous to the economic system". Here one of the major distinctions between the "Induced Innovation Theory" and other models of economic growth and technical change becomes visible. Other theories treat technical change as exogenous to the system in which firms or farmers have to take their resource allocation decisions. Advances in science and technology are thus seen as occurring apart from changes in other socio-economic variables and technology user demand. Therefore, it is difficult to understand the direction technical change takes under different resource and institutional endowments. Following J. R. Hicks, Hayami and Ruttan interpret the majority of technical advances as occurring due to the search of economic agents (farmers) for more profitable technical solutions to their resource constraints and, connected with this, the development and supply of corresponding practices through inventors and private and public technology suppliers. Thus changes in production factor scarcities and relative factor prices result in different technological options being the most profitable ones under differing conditions.

If the majority of farmers in an area is facing similar resource constraints, a bias in farmers' search for new technologies is the result. If new practices are offered to farmers that correspond to this bias technical change in the long-run might take a certain direction, leading for example to an increased development of labor-saving techniques versus land-saving practices. Taking J. R. Hicks' definition of "labor-saving" versus "land-saving" technologies, Hayami and Ruttan (1985) define mechanical technology as "labor-saving" and biological/chemical technology as "land-saving" in the agricultural context. Mechanical technologies, such as tractors, allow substituting power and machinery for labor, resulting in a larger area of land that can be cultivated per worker. Biological/chemical innovations, e.g. high-yielding varieties, fertilizer, conservation practices to restore soil fertility, can substitute land with labor and/or industrial inputs.

Figure 3-1 shows the inducement mechanism for a mechanical (left panel) and a biological (right panel) technology. Line ***BB*** in the left panel and line ***bb*** in the right panel are both isocost lines in period t, in the first case the labor to land ratio and in the second case for the ratio of land to fertilizer use. Point ***P*** in each panel marks the best possible combination of relevant production factors embodied in a certain technology, which corresponds best to the specific economic circumstances. If the relative prices between the production factors change with

Figure 3-1: Induced technical change for mechanical and biological innovations according to the "Induced Innovation Theory"

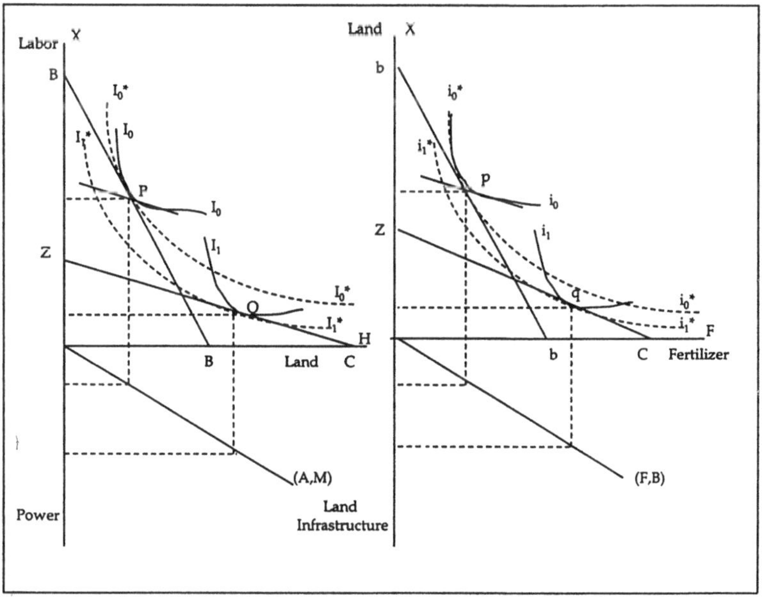

Source: Hayami and Ruttan 1985

time a new isocost line emerges in time t+1, represented as line ***ZC*** in the left panel and line *zc* in the right one. In the case of the left panel, labor prices have risen, while in the right panel land has become more expensive. Answering to these changed economic conditions, the development of a new technology is induced. Point ***Q*** in both panels now depicts the new equilibrium point with the lowest total input costs and the new, corresponding technology. Therefore, farmers operating under these new circumstances would now switch to the new practices if they act according to profit maximizing decision-making rules. Hence, the aim of the implementation of a new technology is to reduce the use of the scarce factor in relation to the other, more abundant inputs. And, quoting again Hayami and Ruttan (1985), "as a result, the constraints on economic growth imposed by resource scarcity are released by technical advances that facilitate the substitution of relatively abundant factors for relatively scarce factors". Nevertheless, this process is not a smooth one. The authors actually state that "in the dynamic process of development, the emergence of imbalances or disequilibria is a critical element in inducing technical change and economic growth" (Hayami and Ruttan 1985, p.92). These imbalances re-focus the attention of technology suppliers on new problems, which are often created by finding a solution to the first set of problems. This then keeps the development process moving.

Figure 3-1 also shows another important feature of the "Induced Innovation Theory". Curves I* represent the so-called "Innovation Possibility Curves" (IPC), I^*_0 at time t and I^*_1 at time t+1. An IPC is "defined as an envelope of unit isoquants corresponding to the alternative technologies that can potentially be developed for a given research budget at a given state of art,..." (Hayami and Ruttan 1985, p. 87) and was first developed by H. Binswanger in his model of a research production function. This curve helps to understand the connection between the technology users (farmers) on the one side and technology developers on the other side. Technology suppliers need to develop technologies that are so attractive to farmers that they are willing to invest in these new practices. Only then can the developers recuperate their research investments. Farmers will only purchase the new technology if it helps to increase profits and with this income, which will only be the case if the new practices help to release farmers' resource constraints. Therefore, two points become visible here: One is how technology developers are induced to produce new technologies that respond to the economic circumstances of technology users. In a later section the technology development side of the model will be investigated in further detail. The second point is how this in return leads to the development of differing mechanical and biological technologies under different economic environments, thus describing the direction of technical advancement in different national economies. Thus, depending on the economic conditions found in a country within a certain point in time, expressed in the isocost line, the process of development and adoption of new technologies takes a particular direction. In an economy where labor is a scarce resource, technological innovations such as tractors that increase labor productivity per unit of land will be created. If land is scarce and therefore in high demand, biological technologies might be of more interest to technology developers and users as they increase output per unit of land.

To empirically test the hypothesis of varying directions of technological change according to differing economic conditions Hayami and Ruttan (1985) compared changes in labor and land productivity and resource prices in different countries. Drawing on results from Japan and the USA, they found that as, for example, land became a relatively scarce resource in Japan, the land to labor costs increased drastically in the period from 1880 to 1980. According to the "Induced Innovation Theory" this in return should have led to a strong research focus on biological technologies that increase land productivity. Empirical results confirmed this hypothesis as agricultural innovations were developed mainly in this field, leading among others to the development of high-yielding varieties for the main staple rice. In the USA land supplies were much more elastic; here labor was in high demand, also due to the high movement of labor to the quickly developing industrial sector. Therefore, technology development focused on mechanical technologies, which provided a rise in labor productivity per worker.

The authors also looked into differing paths of development for a large number of other countries. Comparing agricultural production and productivity changes from 1960 to 1980 in 44 countries, they found three basically distinct curves of agricultural growth. As can be seen in Figure 3-2 they all take the same direction but differ in shape and location depending on initial endowments with land and labor. Lines ***A/L*** describe the land to labor ratios prevailing in the different continents. In the new continents and Libya and South Africa land supply was quite elastic (A/L=100), while the main constraint consisted in the availability of labor. This in return led to a substitution of machinery and power for labor and thus a strong increase in area cultivated per worker. In Asia, Egypt and

Figure 3-2: Differing paths of changes in land and labor productivity between 1960 and 1980 in the agricultural sector

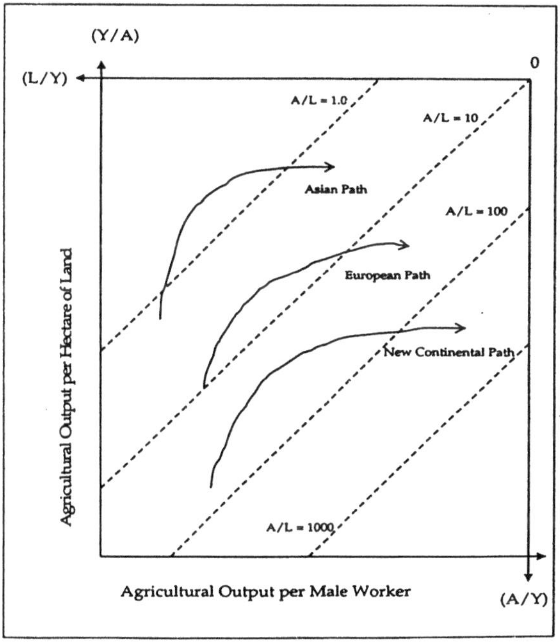

Source: Hayami and Ruttan 1985

Mauritius the situation was the opposite and very rapid population growth since the 1920s, which was not accompanied by a similarly strong development of the industrial sector capable of absorbing the labor, led to a declining man-land ratio. Here the substitution of artificial inputs, such as fertilizer, resulted in a strong rise in agricultural output per unit of land, which is shown by the higher starting point and position of the Asian curve. Europe takes a middle position between Asia and the new continents. For Africa Hayami and Ruttan could not develop the corresponding curve due to a lack in available data.

Therefore, as we have seen the "Induced Innovation Theory" describes a specific mechanism for explaining the switch of farmers to a new technology and the incentives for research institutions in response to changes in relative prices of endowed resources. And we have seen what consequences these differences in relative prices in different regions can have for the direction of technical change. In the following we will investigate if the mechanisms described can also be used to explain the use of EIs.

3.2.2 Environmental innovations and the inducement mechanism described in the "Induced Innovation Theory"

According to the "Induced Innovation Theory", the kind of technologies that are attractive to farmers is determined by their resource endowments, leading them to save first the factor that is the scarcest and thus the most expensive for them. Staying in this framework three points can be considered with respect to EIs.: (*i*) EIs can be analyzed in the same way as CIs, although then some of the most important characteristics of EIs might be overlooked. (*ii*) EIs can be investigated by considering long versus short-term profitability. (*iii*) As EIs are designed to maintain or establish functioning, sustainable agro-ecological interactions, their adoption might better be investigated from this perspective, which then leads to exploring the idea of the 'Productive Capacity of natural resources' as a separate production factor.

(i) EIs treated equally to CIs: One way of understanding EI implementation and development is to place them within the Theory in the same manner as it is done with CIs. In this case, the contribution of a particular EI, such as a specific soil conservation practice, to saving one of the three production factors land, labor or capital will have to be examined. Different EIs can be classified as being more labor or more capital intensive per unit of land. Taking again as an example soil conservation practices, Zurek and Sain (forthcoming) looked into the economic characteristics of the most important soil conservation techniques promoted in Central America (see also Chapter 2). Here it became clear that

most physical structures require a lot of labor for their construction and in most cases additional capital to buy the construction materials and to pay necessary labor. Changes in soil management practices in contrast usually need much less additional labor, their use might simply distribute labor differently within the management system or even reduce its amount. Also, capital requirements are usually relatively small as these technologies in many cases use material available in the area (e.g. for live barriers or cover crops), if they need any at all (e.g. no-burn practices with crop residue management). Depending on their circumstances farmers will have certain biases when examining these new practices. Smallholders in Central America, for example, are usually endowed with a lot of family labor while capital availability constitutes an important constraint, as their income is quite low and credit can only be obtained with secure land titles difficult to acquire for most subsistence farmers (see also Chapter 2). Under these conditions it is understandable that farmers are more inclined to use changed soil management practices than physical structures. But, as the low adoption level of many promoted soil conservation practices shows, when compared with the traditional farming practices, few of the new techniques fulfill any of the requirements laid out in the "Induced Innovation Theory" to explain the switch of farmers in technologies and its direction. They might not contribute enough to reduce total input costs. And they are likely not to save on the scarce capital factor. Under these premises the implementation of these new EIs does not make economic sense to farmers, at least in the short run. An example of a successful conservation practice, whose adoption worked according to the Theory, is the use of Zero and Minimum Tillage systems in Azuero, Panama (Pereira de Herrera and Sain 1999). The main driving force behind farmers' use of these new tillage systems, which are good erosion control measures, was the reduction in costs of farm operations. In these systems fewer or no plowing is needed, so costs for machinery and fuel declined, which induced farmers to switch to the new practices.

If the only way of analysis for EIs though is the one similar to CIs, a fundamental point of EIs is missed: Their main purpose is to maintain or establish functioning, sustainable agro-ecological interactions and thus contribute to the reduction of environmental degradation. The extent to which this purpose is achieved by the investigated practice has to be included in the analysis in one way or another. On the other side, it also becomes clear that if only this particular aspect of environmentally sound practices is emphasized, but their economic characteristics are not examined at the same time, it is not very likely that these technologies will be adopted by farmers as they will not fit most farmers' circumstances.

(ii) Short versus long-term profitability aspects: A second way of explaining farmers' behavior with respect to EIs within the framework of the "Induced Innovation Theory" is to look at one of the important distinctions between CIs and EIs mentioned earlier, namely the short- and long-term profitability aspects connected with both technology groups. While CIs are designed to increase profitability of farming activities in the short-run, EIs become profitable in most cases only after an extended period of time. If degradation occurs on his fields, the farmer faces different possible yield development paths depending on the choice he makes between continuing or changing his degrading farming practices (Pagiola 1994). Figure 3-3 demonstrates these possible paths for the case of soil erosion. Path A shows how yields decline substantially with time if the soil degrading techniques are continued. How strong this decline will be depends on a large number of ecological and management factors (see also Chapter 2). In path B the use of inputs and practices that are known to cause erosion is reduced (e.g. fields are not burned as often), which results in a recovery of yields after some

Figure 3-3: Possible paths of yield development depending on application of different farm management practices

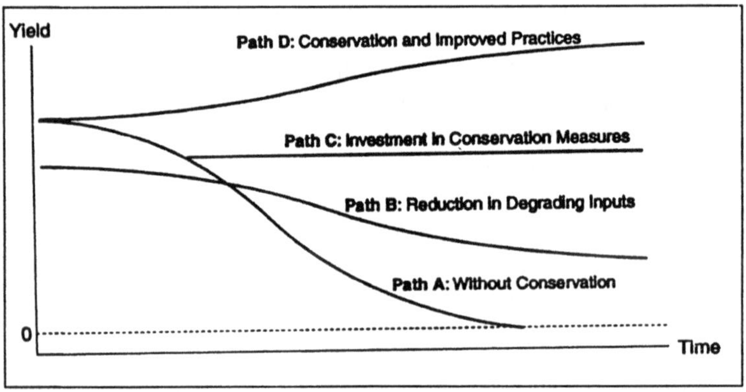

Source: Pagiola 1994

time, although they do not return to their initial level completely. Paths C and D show different degrees of investment in conservation and additional improved management practices. Depending on the measures taken, yields will recover and can even rise again. The decisive question here is how fast this recovery process takes place and when first results become visible, which is influenced by the kind of conservation measures and improved practices implemented. According to their perceptions on the time lag between incurred costs and expected future benefits and on the amount of received benefits, farmers will have to decide how high current costs can be and how long they are willing to forego profits. As described earlier, under subsistence conditions farmers are likely to have very high discount rates. This in return reduces the likelihood that small-scale farmers perceive many EIs as being more profitable for them than their traditional farming practices. This specific characteristic of EIs places them behind CIs in the priority list of subsistence farmers. There are two ways for this prioritization to change:

1) The EI includes in addition some means to raise productivity in a relatively short period. These practices are then productivity-enhancing, resource-conserving (PERC) technologies, as they posses another characteristic than just the conservation side that makes them attractive to technology users. Vosti and Reardon (1997) also call these kinds of practices 'Overlap Technologies', as they are designed to address sustainability issues together with the need to sustain economic growth through productivity increases. The authors conclude that the proper design of these technologies is not easy, "....but no substitute strategy is appropriate." (Vosti and Reardon 1997, p. 11). A PERC or Overlap-Technology can thus be seen either as an EI with a productivity increasing aspect or as a CIs that has some degradation reducing properties. Which one of the two aspects is regarded as the more important one depends on the perspective of the technology user. For subsistence farmers definitely the CI-properties of the PERC technologies are more important. With these technologies the "Induced Innovation Theory" inducement mechanism works in the same way as for "normal" CIs, as they are able to address the short-term profitability demands of farmers. In Central America at least two sets of practices are available to farmers in different areas that address both conservation and productivity aspects. These are No-burn systems with crop residue management and Legume Inter-cropping/ Cover crop systems. For the purpose of this study the use of the leguminous cover crop Mucuna was investigated in more detail. As the legume provides nitrogen to the main crop, it is able to raise the yield of the main crop substantially in most cases within one or two cropping seasons, while reducing soil erosion significantly by providing a thorough cover to the soil. In a case study from Guatemala, which will be presented in Chapter 4, farmers' perceptions and the factors influencing the use of this legume are studied.
2) Another option to change farmers priority list can be to try and change farmers' time preferences. If farmers begin to place a value on the importance of well functioning agro-ecological interactions they might be inclined to consider the possibility of foregoing profits now to gain more profits in the

future. For this to happen farmers will have to be aware of the exact relationships between environmental degradation and farm output. Thus farmers need the corresponding background information. An overall higher education level would help in this respect. Nevertheless, if the conditions in which they have to take this decision do not allow farmers to lose any of the gains in the current period they are still not very likely to opt for this possibility. This problem can be particularly observed under subsistence conditions. In this case the provision of knowledge will have to be coupled with adjusting farmers' circumstances, such as land tenure or credit facilities.

In both described cases the decision of farmers to switch from the traditional farming practices to an EI follows the inducement mechanism laid out in the "Induced Innovation Theory". A comparison of profitability levels between the new and old technology will either lead to a change in practices or not. As low adoption levels of many conservation practices demonstrate for the case of soil erosion, EIs often lose in this kind of comparison. But nevertheless there are a few options in technology design and by trying to change some farmers' preferences that could be employed to manipulate this outcome.

(iii) "Productive capacity" as a new production factor: A third way of interpreting EIs implementation within the "Induced Innovation Theory" framework is to go back to the original reason for EI development: to maintain or establish functioning, sustainable agro-ecological interactions and to contribute in this way to the reduction of environmental degradation. Here nature with its support function for agricultural production is the main focus. Functioning interactions between the natural ecosystem and the agricultural production system are to be created and sustained. Therefore the question arises whether these interactions can be interpreted as a factor influencing the production output in the same way as the neoclassical production factors labor, capital or land, which all determine the production function of a particular agricultural system. Similar to the other production factors a farmer has to raise the input of these agro-ecological interactions, thus their number and quality, in order to reach a higher production output. In this interpretation profitability of the farmer's use of inputs becomes, at least in the long run, explicitly dependent also on the complementarity of these interactions. Output is reduced as the degree of functioning relationships declines, if the other production factors are not substituted for it. Hence, for example, a farmer might have to invest in fertilizer and thus use capital to maintain soil fertility. Then the kind of technological options that help the farmer to avoid this decline, namely EIs, become appealing. If we hypothesize that we are able to name 'Productive capacity' as an additional production factor, it might be possible as well to analyze these relationships within the "Induced Innovation

Theory" framework. EIs are able to save on the factor 'Productive capacity' by reducing the misuse of nature. If due to the occurrence of environmental degradation, agro-ecological interactions deteriorate, they might become a scarce, sometimes even a decisive, factor in the production function. According to the Theory, technologies such as EIs that save on the scarcest factor, and most expensive one, to the farmer have then a higher potential for adoption by farmers than other technological options.

Figure 3-4 illustrates the described relationships within the Theory's framework. The Isocost line at time t represents different combinations of the new production factor 'Productive capacity' and one of the traditional production factors labor, capital or land that all result in the same amount of input costs. Isoquant CI depicts different combinations of 'Productive capacity' and one of the other production factors, embodied in a certain commercial technology that will provide the same output level. At point ***e*** the best ratio of these two production factors is reached. A well functioning, economically efficient farm will operate at this optimum point with the corresponding CI. Over a certain period of time and through the misuse of the 'Productive capacity' that could result from the use of the employed CI-technology, the quality of this production factor deteriorates.

Figure 3-4: The inducement mechanism for the use of environmental innovations with 'Productive Capacity' as a new production factor

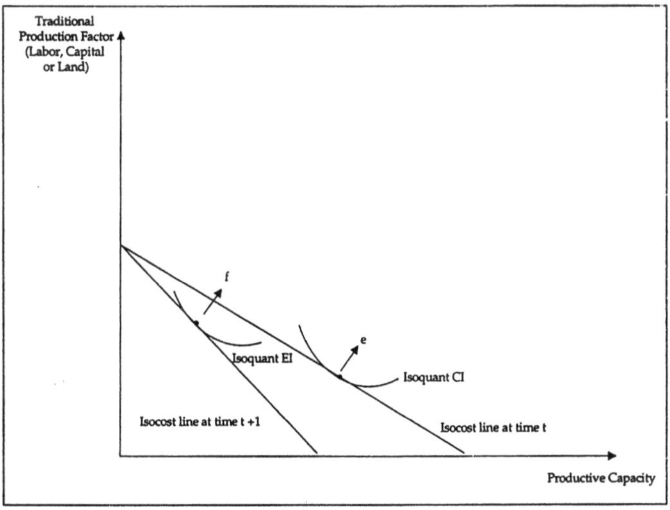

Source: Own graph

As an example, one can think of the use of burning of crop residue practices before planting, which is still commonly used in Central America. If fallow periods are reduced too much the continuing use of burning practices can lead to severe erosion problems, resulting in deteriorating soil fertility. The loss of soil fertility indicates the reduction in the number and quality of agro-ecological interactions necessary to maintain agricultural production. Thus the use of the 'Productive capacity' becomes more difficult and with this also more expensive for the farmer. He might have to substitute one of the other production factors for 'Productive capacity', e.g. invest capital to buy fertilizer. This then results in a new Isocost line at time t+1 with a different slope as Isocost line t, which indicates that less 'Productive capacity' can be used. Under the new circumstances a technology different from the one used before can become attractive to the farmer as this new practice utilizes less 'Productive capacity' than the CI employed before. This new technique, an EI, might provide the same output level as the CI, thus show the same isoquant as the CI. But this output is now achieved using a lower level of inputs, especially of the 'Productive capacity'. Assuming again that the farmer will take his management decisions according to a profit-maximizing rationale, he will adopt the new technology. The farm will now operate at the new optimum point *f*.

Therefore, if we can hypothesize about a new production factor that captures the existence of sustainable agro-ecological interactions in the agricultural production system, the use of EIs can make economic sense to the farmer. Under the assumption that farmers' time horizons are long enough, it can then also be hypothesized that farmers in more marginal production areas, such as hillsides or semi-arid areas where the occurrence of degradation is usually visible much faster than in favorable agro-ecological environments, are more interested in EIs than farmers in other regions.

For its practical implementation though, the interpretation of the 'Productive capacity' as a production factor carries one decisive problem: It is difficult to assign a monetary value to the 'Productive capacity'. What does it cost for example to have a reduction in the water holding capacity of a particular soil due to a decline in its organic matter content as a result of soil erosion? As organic matter content is part of a complex system of relationships within the soil and each soil reacts differently, it is difficult to really assess the impact of its loss thoroughly. Or what is the price of loosing a certain degree of biodiversity within the agro-ecosystem? Some kind of a valuation is needed for two reasons: First, if a monetary value exists for valuing these interactions and they could be assigned some kind of price, these interactions would be comparable with the other production factors. It would then be possible to include them in farmers' calculations of his production expenses. And farmers could then also calculate the expense of substituting them for another production factor and assess their importance for the production process. Attempts to find a proxi variable with respect to evaluating the importance of functioning agro-ecological interactions have been made by using natural resource accounting models (Magrath 1998, Bishop and Allen 1989) or calculating for example the replacement costs of lost soil nutrients. Second, assuming that the relative price mechanism described by Hayami and Ruttan (1985) for determining the direction of technical change and giving signals to technology suppliers when developing new technology options works correctly, some kind of valuation for functioning agro-ecological interactions, would make it possible and/or easier to include them in the considerations of developers. This facilitates the tailoring of new technologies with stronger or weaker EI characteristics to farmers' constraints, depending on the severity of the degradation problem the farmer faces. So far though the monetary valuation of the 'Productive Capacity' is a difficult question. And as long as there is no real progress made in this direction the price mechanism laid out in the Theory is difficult to employ and we will have to think of other ways to effectively guide technology development and the interactions between technology users and suppliers. In the next section this problem will be investigated in more detail.

3.2.3 Technology development according to the "Induced Innovation Theory"

From the earlier explanations of the "Induced Innovation Theory" it is obvious that farmers' profit maximizing behavior induces the use of new practices and thus leads to technical change by evoking a continuous search for ways to increase net benefits through a more profitable allocation of resources. Another pivotal point of the theory has become clear as well: For this process to take place new technologies have to be developed continuously and offered to farmers, resulting in an Innovation Possibility Curve (IPC) that represents the potential technical solutions to particular production factor constraints of farmers. Thus farmers' demand for new practices induces technology suppliers, ranging from the public agricultural research institutions to the private technology developers and input supply firms, to invest in research and technology development as they hope to recuperate their costs when farmers purchase the more profitable technologies. But this also implies that technology developers are able to take up the signals reflecting resource constraints of farmers and develop effective solutions. According to Hayami and Ruttan (1985, p.87) "in the Binswanger model, technical change is guided along the IPC by changes in relative factor prices, while the IPC itself is induced to shift inward toward the origin by the growth in product demand". This process though only happens in an efficient way, 1) if prices are not distorted and represent changes in input and output markets correctly and 2) if there is an active exchange of information and a high degree of interaction between farmers and technology developers. Differences in perceptions between farmers and researchers concerning input or output prices can lead to the development of technologies with a low adoption potential. In the following we will investigate if this might be one of the explanations for understanding the low adoption levels of some of the promoted soil conservation practices in Central America.

Concerning the interactions between farmers and researchers, Hayami and Ruttan (1985) suggest that these interactions can take place most effectively if farmers are organized into groups, while especially publicly funded agricultural research should be decentralized, with different experimental stations etc. in various regions. For the whole "Induced Innovation Model" the efficient and correct responses of technology developers to farmers' search for new technologies represent the critical point for the inducement mechanism to work. If this response fails, a slowing-down in the whole development process will be the consequence.

Hayami and Ruttan (1985) explicitly include public organizations as technology providers in their model. Different from private technology developers or input suppliers, these organizations do not necessarily have the incentive to respond closely to farmers' demand and price signals of input and output markets as their funding does not depend on the recuperation of research investments from technology users. Nevertheless, the authors propose that a well-functioning system, which offers professional recognition and rewards to researchers, could provide incentives to guide scientists and research administrators in their research priorities to find solutions to the most pressing problems of society. In this way not only efficient applied research would be undertaken, but also researchers in the more basic sciences would be induced to respond to research questions arising from the applied field. This in return leads to an efficient public R&D system within a country, necessary to sustain agricultural productivity growth. It should be noted, as Hayami and Ruttan (1985) also describe, that the fast development of the agricultural sector in most developed nations has been and still is closely related to efficient, publicly funded research organizations that have responded quickly to farmers' most important resource allocation problems. They additionally explain the importance of these organizations for the provision of public goods, whose development is likely not to be attractive for private technology suppliers. As these goods are characterized by non-rivalry and non-excludability, private firms will have difficulties in protecting the results of their research investments or in capturing their benefits. Public institutions can here play an important role in supplying basic research results and in catering to groups of technology users of less interest to the private firms, such as small-scale farmers for example.

A last important aspect of technical change should be mentioned here: Hayami and Ruttan (1985) do not suggest that technical change is always guided through the described inducement mechanism. They acknowledge the important contribution to technological innovation arising from the advances in general knowledge and sciences. Thus new solutions that can reduce the costs of the innovation process could have contributed to technical change without being influenced by changes in relative prices of inputs or changed product demand.

As we have seen, the "Induced Innovation Theory" proposes a framework for understanding the interactions between researchers and farmers necessary to guide a targeted and effective technology development process. In the following we will investigate whether the described mechanisms also apply to the development of EIs and if the low adoption levels of EIs in Central America can be understood from a technology development side as well.

3.2.4 Interactions between technology users and suppliers with respect to environmental innovations

As described in an earlier section, EIs aim at saving or restoring functioning, sustainable agro-ecological interactions. But so far it is difficult to evaluate these interactions, or the 'Productive capacity' of natural resources, in the same way as the neo-classical production factors, though they can influence agricultural production in a similar way. But what does this mean when trying to understand the mechanisms that are needed to guide EI development along an efficient path?

In the case of CIs development the "Induced Innovation Theory" describes a mechanism for effectively guiding the interactions between technology users and suppliers. Relative prices of production factors, which result from relative factor scarcities, signal farmers' priorities to technology developers when looking for new practices that not only increase productivity or reduce input costs but also save on the scarcest factor. For EIs though, it is difficult to apply this mechanism per se as their aim is to save on the 'Productive capacity' by reducing environmental degradation. But, as described earlier, it is difficult to assign a monetary value to this characteristic, which in return makes it difficult to really compare its importance for the farmer with the importance of other factors. The price mechanism signaling farmers' resource scarcities thus cannot work here. Therefore, even if a farmer faces environmental degradation on his farm it is at first difficult for him to assess what this really means for him in terms of a decline in income. And second, technology developers will have difficulties estimating the farmer's need and interest in implementing a technology with explicit EI characteristics. Therefore, farmers' and researchers' perception of the value assigned to the 'Productive Capacity' can differ substantially, which in return might lead scientists to develop practices that do not correspond with the resource constraints as perceived by the farmer. Going back to Figure 3-4, the isocost line at time t can be also interpreted as the isocost line the farmer perceives. The use of the production factor 'Productive Capacity' is here not as expensive and with this as restricted as with the isocost line at time t +1, which is the isocost line the researcher might see. Figure 3-5 illustrates this new interpretation. The difference

Figure 3-5: Different perceptions of farmers and researcher concerning the Isocost lines depicting the price ratio between the 'Productive Capacity' and another production factor

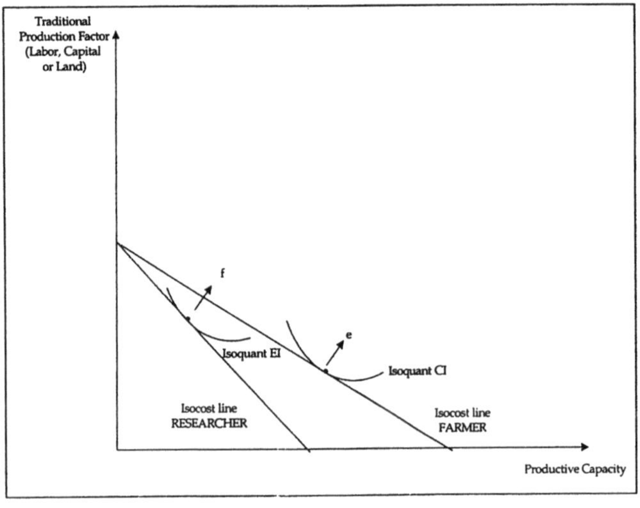

Source: Own graph

in perception between farmers and researchers can arise from different reasons. The scientist might calculate a higher price for the 'Productive Capacity' as he/she sees the consequences of environmental degradation for agricultural production as being more severe than the farmer, especially in the long run. Depending on the farmer's institutional setting, and especially if land property rights are unclear, his time horizon is likely to differ from the scientist perspective. In addition, the researcher might include the externalities of nature degradation, i.e. the costs of degradation for society, in his valuation of the 'Productive capacity'. The scientist's analysis of prices will lead him to develop an EI that saves on the scarce factor 'Productive capacity'. The farmer in his perceived production environment though will have no incentive to switch to the EI, but rather continue using a CI.

The different perceptions of farmers and researchers can also be illustrated in another way. Figure 3-6 shows again two different isocost lines as they are perceived by the farmer and by the researcher. This time though the isocost lines reflect the ratio between the price of the 'Productive capacity' at a certain point in time, called t, and a later point in time, called t+1. For the researcher, who is interested in the long-term, sustainable use of nature and likes to avoid environmental degradation as much as possible, the 'Productive capacity' of natural resources has to be used as careful as possible and therefore this production factor at time t is perceived as scarcer and thus more expensive than for the farmer. Through this behavior the scientist expects to be able to continue the use of the 'Productive capacity' at time t+1. Especially subsistence farmers though usually have a shorter-term vision of 'Productive capacity' and often have to exploit agro-ecological relationships as much as possible to obtain sufficient food in order to guarantee the survival of their families. Therefore they are likely to have a discount rate differing from the researcher's and do not necessarily include the consideration of the future state of the resource they are using into their calculations. This might be especially the case with land-renters as they might not be able to farm the same plot of land in the future. For these reasons farmers might assign a lower value to the production factor 'Productive capacity' as researchers. Fitting an Innovation Possibility Curve to these isocost lines could then help to understand the low adoption levels of certain EIs. If environmentally sound practices are developed by scientists in response to isocost line RESEARCHER in Figure 3-6 they might not fit farmers perception of the value of this production factor. Hence, the switch from the CIs that might fit isocost line FARMERS does not make sense to the technology users.

As these different perceptions of farmers and researcher concerning the value of the productive capacity of natural resources exist and this can lead to the development of conservation practices with a low adoption potential the question arises if another kind of mechanism is required to guide technology development. A few points should be mentioned here concerning characteristics of this process:

As already proposed by Hayami and Ruttan (1985), there has to be an active and vivid exchange of information between farmers and researchers so that researchers are able to really comprehend farmers' constraints and their strategies to cope with these constraints. This will also let them understand farmers' priority list of problems. As natural resource degradation problems and also resource

Figure 3-6: Different perceptions of farmers and researcher concerning the use of the production factor 'Productive capacity'

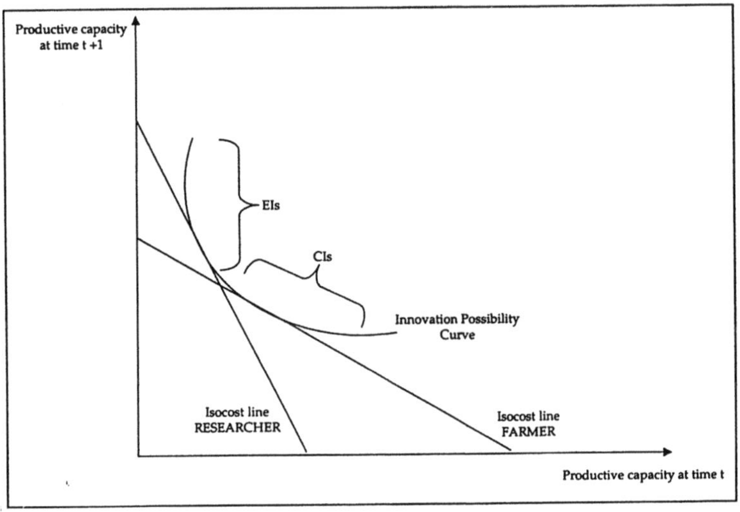

Source: Own graph

endowments are often embedded in a complex, location-specific set of interactions, their solutions will also have to be location-specific. Developing mechanisms to cope with this characteristic have already been addressed in On-Farm Research and in the development of participatory research methods. But there is probably further development and application of these new methodologies needed to properly elicit farmers' demand. Additionally, these methodologies often require a certain degree of self-organization of farmers with similar demands.

And another question arises here: Are suppliers of EIs aware that they need to employ new mechanisms to really develop the technologies that answer farmers' resource degradation problems while addressing some of their other resource endowment constraints as well? Do they see that the differences in perceptions of the value of the production factor 'Productive capacity'? What kind of organizations works at all in the development and promotion of EIs? And how do they select the technologies they promote and/or develop further with farmers? In the second case study carried out for this thesis and presented in Chapter 5 some of these questions will be investigated. Taking a county in El Salvador as an example the organizations that directly work together with farmers in the promotion and development of soil conservation technologies are studied. In addition, farmers' response, namely the adoption level of the promoted conservation practices, and their information sources for these practices are investigated.

3.2.5 Induced institutional change and environmental innovations

In the "Induced Innovation Theory" Hayami and Ruttan (1985) not only investigate sources and direction of technical change but also propose hypotheses on how institutional change occurs within a society. Similar to technological change, they also view changes in institutions as endogenous to the overall economic development process.

For the authors "institutions are the rules of a society or of organizations that facilitate coordination among people by helping them form expectations which each person can reasonably hold in dealing with others" (Hayami and Ruttan 1985, p.94). Thus they regulate human behavior and facilitate, for example in the economic context, the formation of expectations on resource use and the distribution of income. Institutions need to have two basic characteristics. On the one side they have to be steady over a long enough time period for people to be able to work with and to influence decisions. On the other side they also should be open to change necessary to facilitate the development process.

Hayami and Ruttan (1985) describe two sources of demand for changes in institutional settings within a society: 1) changes in resource availability and 2) technical change. Changes in relative prices of production factors are an indicator of resource availability modifications within a society. These alterations can in return induce the need to develop new institutions. Changes in land use and property rights in different nations can be used as a good example. The authors take here the development of land property rights in Japan to demonstrate how population growth led to land scarcity and land use intensification, which then resulted in new lease and mortgaging arrangements for land among peasants. These de facto new land use rights later served as a basis for the modern property right distribution in the last Japanese empire. Institutional change can also be induced by technical innovation as the development of high-yielding varieties has shown.

As we have seen before (section 3.1), an important feature of EIs is that their successful implementation often requires the collaborative effort of a group of farmers or a community to reach a measurable impact in reducing environmental degradation. This implies a certain degree of self-organization and communication among farmers. In the long-run this can lead to changes in the social infrastructure of a community and to establishing institutional changes as described by Hayami and Ruttan. Additionally, many EIs need a certain period of time until their positive impact on agricultural production can be seen. This implies that farmers will only have an incentive to implement these practices if either relatively stable and secure land use rights exist or if all farmers in the area invest in the same way into maintaining soil fertility. In this case there has to exist a general agreement amongst farmers on the importance of soil fertility maintenance, which can then be interpreted as a new institutional rule according to the Theory. One example for the existence of such an agreement can be found in the case of the adoption of No burn practices with crop residue management in Guaymango, El Salvador, which is well documented by Sain and Barreto (1996).

According to Hayami and Ruttan (1985) the supply sources of institutional change have so far not been studied well. They state additionally that "... the supply of institutional innovations is strongly influenced by the cost of achieving social consensus (or of suppressing opposition). How costly a form of institutional change is to be accepted in a society depends on the power structure among vested interest groups" (Hayami and Ruttan 1985, p. 96). Here the social and cultural aspect of institutions and their connection to the political organization of a nation becomes clear. Similarly to the influence that progress in general natural sciences can have on technical change the authors propose that social science advances and education can have a positive impact on reducing the costs of developing new institutions. Nevertheless, in this field important research questions are still open, such as the question if and how it is possible to close the gap between private gains of politicians from the development of certain new institutional types and benefits for society as a whole arising from these innovations. As in this thesis the main focus is put on the adoption mechanisms of new environmental technologies and the relationship between technology users and suppliers, institutional change due to the implementation of EIs is not investigated further.

3.3 Summary

In the "Induced Innovation Theory" Hayami and Ruttan (1985) developed a framework to explain technical and institutional change and its direction in the agricultural sector. New technologies that can be successfully adopted are developed in response to the resource endowments of technology users, i.e. farmers. Technical progress and changes in resource availability result in the need for new institutional settings, inducing institutional change. The cost of achieving new institutions is influenced by cultural variables and the political structure of a nation, while new social rules and regulations in return can have an impact on cultural practices, technology use and resource availability. It is important to note here that all four variables are not only connected with each other but that their interactions are described as recursive. Therefore, any alteration in one of the variables induces changes in the other factors as well. In addition, if the interaction cycles are not closed, e.g. if there is no active exchange between technology users and developers on resource availabilities and relative prices of these resources, the innovation process cannot take place and development is impeded.

In this chapter the mechanisms laid out in the Theory for commercial innovations were investigated for their applicability to environmental innovations, which possess some specific characteristics different from CIs. Here mainly the relationship between the perception of farmers and technology suppliers on resource endowments and the consequences of this for technology development and adoption were discussed. As differences in the perceptions of farmers and researchers on the value of the 'Productive capacity' of natural resources exist and the price mechanism described in the Theory for guiding technology development is applicable only with difficulty, the effective feedback loops do not work. Therefore, the process of technical change to answer to the resource degradation problem is stagnant. If the process of technical change is not or only very slowly taking place, changes in the institutional framework are unlikely to occur. Therefore, governments and organizations concerned about natural resource degradation have to think of additional effective technology design and incentive mechanisms for farmers to foster the widespread development and adoption of environmentally sound farming practices.

4 The Adoption of Productivity-Enhancing, Resource-Conserving (PERC) Technologies in Central America: The Case of the Legume Mucuna (*Mucuna spec.*) in the Polochic Valley, Guatemala

In comparison to Commercial Innovations (CIs), the adoption of Environmental Innovations (EIs) by farmers has often faced additional obstacles. This is particularly the case in smallholder agricultural systems. Small-scale farmers are confronted with a number of constraints when searching for new technologies, such as information deficits, capital limitations, insecure land titles, and weak institutional settings. Under such conditions it is even more difficult to see for farmers why they should switch from traditional farming practices to more resource conserving techniques, whose economic benefits are visible only in the long-run. These specific characteristics of EIs, as explained in the preceding chapter, place them behind CIs in the farmers' priority list and partly explain the unsatisfactory adoption rates found not only in Central America. Productivity-enhancing, resource conserving (PERC) technologies are seen as one possible solution as they give farmers an additional incentive for adoption by helping to increase yields in a relatively short period of time.

This situation raises some important questions: Which aspects are important for subsistence farmers in Central America when choosing a new technology, especially if soil degradation is a visible problem? How can their behavior be modeled when a PERC technology is involved, which covers both aspects of productivity increase and conservation? To find some answers to these questions a case study was carried out in an agricultural frontier area in Central America, where small-scale farmers are the majority of land users and soil erosion is an eminent problem.

4.1 Introduction

In the following a case study will be presented, which was carried out in the Polochic Valley, Guatemala, on the use of the legume Mucuna/Velvetbean (*Mucuna spec.*) as a cover crop intercropped with maize. This practice is considered to be a PERC technology as Mucuna not only helps to boost maize yields by supplying important plant nutrients to the crop, but is also seen as an effective soil cover, thus reducing exposure of soil to erosion on hillsides. In collaboration the International Maize and Wheat Improvement Center (CIMMYT) and the Central American maize scientist network *Programa Regional de Maíz* (PRM) have worked for the last ten years with PERC technologies for maize based cropping systems in the region. This study is part of the effort of CIMMYT/PRM to evaluate these technologies from an economic point of view, and to investigate factors that influence adoption by smallholders, the main maize producers in the region. The study was carried out in the Polochic Valley, an agricultural frontier area in northeastern Guatemala. There the system of intercropping Mucuna with maize is widely known and used by a large number of farmers since it was introduced in the 1930's.

The study presented here has the following specific objectives:

- To describe the Maize Production System of small-scale farmers in the Polochic Valley.
- To investigate the current adoption rate of a PERC technology, namely the Maize-Mucuna-System, by small-scale farmers in the valley.
- To describe the interactions between the economic, social and institutional factors that lead to the use, non-use or abandonment of this system.
- To investigate the perception of farmers of soil erosion and its effects and their use of soil conserving technologies.

The results of this study shall contribute to answer the following questions:

- Why do farmers use a PERC Technology – for its productivity-enhancing or its resource-conserving effects, or for both?
- How can we model the motivation behind farmers' technology choice?
- What are the factors that shape farmers' intention?
- What does this mean for technology development and policy design to promote the use of PERC technologies?

4.2 Methodological issues

In March/April 1998 a survey of 137 maize producing farmers was conducted in the county of Panzós, the central county of the valley. The survey was carried out in collaboration with the personnel of the ICTA (Guatemalan National Agricultural Research Organization), the former DIGESA (Guatemalan public extension service) extension workers, and translators for Kek'chi of the NGO CARE International. Farmers were interviewed (reference year 1997) about the following topics (survey in Annex 1):

- Farm assets and cropping system
- Use of Mucuna and the reasons for its use, non-use or abandonment
- Knowledge about soil erosion and use of soil conservation methods

- Maize production system in either the first or the second cropping season on the biggest maize plot (two data subsets: 63 farmers for the first season, 74 for the second season)
- Credit market
- Labor distribution among family members
- Information sources and participation in community organizations

The data were analyzed using first a Logit model to assess the likelihood of farmers with certain characteristics to use the Maize-Mucuna-System. In this way the decision of the farmers using measurable variables, like farm size, age or land tenancy, can be modeled. This kind of analysis however does not allow the inclusion of certain unobservable factors, like perceptions, knowledge, opinions or believes about certain subjects, that might play an important role in the final decision of the farmer to use a certain technology. They might be especially significant in the decision to use conservation technologies. As investigated in Chapter 3, benefits of Environmental Innovations are usually seen only after a substantial period of time. With a Logit model it is difficult to capture these long- versus short-term profitability aspects, which are so important to understand farmers' adoption behavior. In addition, the adoption of environmental innovations might be even more beneficial to society than to the individual farmer (see also Chapter 3). This is why it is important to find a way to include certain hidden factors, which are difficult to measure, in the analysis of the adoption decision.

One way in which hidden, underlying factors can be analyzed is through a Structural Equation Model with latent variables. This type of model is often used in marketing research, psychology and social sciences. It uses latent variables to represent unobservable factors, whose causal relationships among each other are modeled. Each of the latent variables can be assigned a number of measurable indicators, which all together define the variable. A regression analysis is run between these variables to assess the strength of their relationships. Thus in such models an external information structure of measurable indicators and an internal causality structure of latent variables exists. Both the Logit and the Structural Equation Model will be explained in further detail in Section 4.5.

4.3 Description of the Study Area

The Polochic Valley is considered to be an agricultural frontier area. The more intensive agricultural use of the region only started around the middle of the 20th century. Access to the valley used to be relatively difficult, isolating the valley from the rest of the country. This will probably change with a new road being built in 1999/2000, which is also thought to impact on the valley's population. Also changes in land use are expected as many parts of the valley are seen as high potential areas for agricultural production (L. Chavez and O. Garcia, personal communication 1997).

4.3.1 Agro-ecological conditions

The Polochic Valley lies in northeastern Guatemala in the transition zone between the highlands and the low-lying rainforest areas of the Petén, covering major parts of the Departments of Alta Verapaz and Izabal (total area: 2822 km^2). It stretches westwards from Lake Izabal, Guatemala's biggest lake, and is bordered by two mountain ranges: the Sierra de las Minas in the south and the Sierra de Santa Cruz in the north. The Polochic River, which is fed by many smaller rivers from the mountains, runs eastwards through the valley into the lake (Figure 4-1).

The county of Panzós lies in the middle of the valley. Most of its communities are located between 5 to 300 masl. The area's climate can be characterized as hot-humid subtropical with an annual precipitation of 2500 to 3000mm, an average temperature of 26°C and a relative humidity of about 80%. The rainy season lasts from May to December, with July being the month with the highest precipitation (Figure 4-2) (Chavez 1994, Fournier 2000).

According to the FAO soil classification the soils in the valley can be divided into three classes: First, Eutric Glyesols are the dominant soil type in the area. They are characterized as loamy to clayey, with slight acidic pH, bad drainage and a high soil erosion potential. Second, Dystric Cambisols, which can be found in the northern part of the valley, have a low pH as well, but a better drainage and water holding capacity. Located in the south, Orthic Luvisols have a relatively high fertility, though their high clay content can lead to drainage and erosion problems (Fournier 2000). Soil depth in general is medium to shallow.

In general three agro-ecological zones can be identified in the valley (Garcia 1992):

- Hillsides, which slopes up to 60%
- Dry plains, not flooded
- Flood plains of the Polochic River, which are partially inundated from May to September

Figure 4-1: Map of Guatemala and the study area

Source: University of Texas, map collection

Figure 4-2: Average Maximum, Minimum and Mean Temperatures, Precipitation and Evapotranspiration in Panzós, Guatemala, from 1969 to 1989

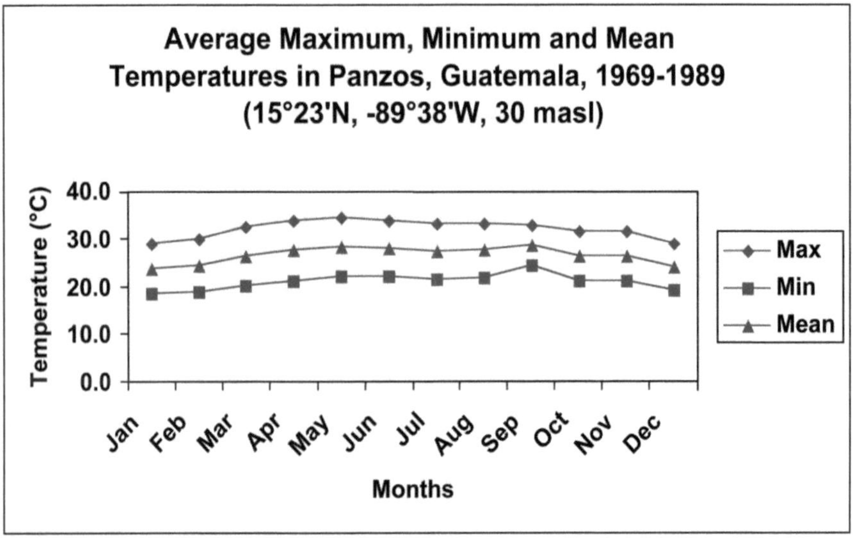

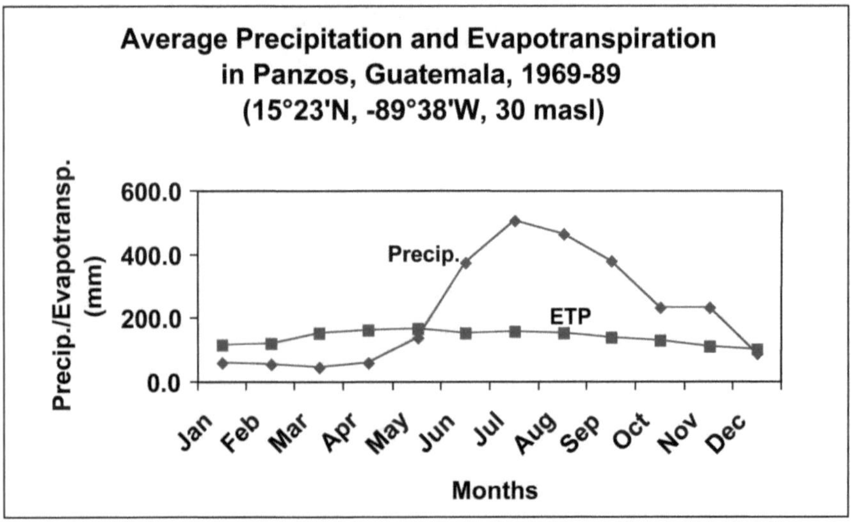

Source: Fournier, unpublished

4.3.2 Socio-economic conditions

The valley's population mainly consists of Kek'chi Mayans (92%), who moved to the Polochic Valley from the highland regions of Alta Verapaz due to population pressure in their home counties. They mostly speak their native language and no Spanish; the illiteracy rate lies at 74% in the municipality of Panzós. In general, population pressure in the valley is not considered high (52 inh./km^2), but the population growth rate is quite elevated (2.9%) (Hernández Jiménez and Silva Goméz 1994; CARE n.d.).

Agriculture is the main source of income for the population (Hernández Jiménez and Silva Goméz 1994). Small-scale farmers farm the hillsides and parts of the dry plains using traditional slash-and-burn methods for subsistence production of maize, beans and some cash crops. Not more than 30% of the production is sold on the market. Middlemen buy surplus production, which is then transported to Guatemala City and other markets outside the valley. Farmers own the land they work on in some kind of formal/informal way but there also exist communally owned fields (Hernández Jiménez and Silva Goméz 1994). Credit is difficult to obtain for most smallholders because they often do not have a proper land title and are not familiar with procedures to obtain credit. In addition, interest rates are relatively high (12% annually at the bank, up to 100% annually at private money lenders) (own investigation 1998).

European and American settlers began to move into the valley since the turn of the 20th century, introducing extensive cattle ranching and commercial farming practices for the cultivation of bananas, coffee, cardamom and some horticultural crops. The United Fruit Company owned extensive tracts of land near the lake until the 1950s, using them for banana plantations (L. Chavez, personal communication 1997). More commercially oriented farms and cattle farms are mainly located within the flood and the dry plains, both providing favourable conditions for agriculture. Here, also rice is produced during the high rainfall period from May to August and medium size farmers often additionally produce tomatoes and melons.

Since 1988 agricultural research is carried out in the Polochic Valley by ICTA, Guatemala's public agricultural research institution. ICTA has a local office in Panzós. Owing to structural changes in ICTA, staff has been reduced, leaving only one researcher for the whole valley. Since DIGESA (Guatemala's public extension service) closed in 1998, only a few NGOs, like CARE International, provide technical assistance leaving the large majority of farmers without assistance. The municipality is supporting a few small programs promoting no-burn practices with crop residue management and the reforestation of water source areas by contracting some of the former public extension workers. Unfortunately, funding for these programs is very limited (A. Villafuerte and J. Cortéz, personal communication 1998).

4.3.3 The general farming system in the Polochic Valley

The main annual crops grown in the valley are maize, beans and rice. They are cultivated primarily by small (less than 8 ha) and medium (between 8 and 100 ha) size farmers. Big farmers rather produce cattle in extensive systems in the dry and the flood plains and coffee, cardamom, and citrus fruits in the hillsides (Chavez 1995). Nevertheless, small-scale farmers often also cultivate coffee and cardamom as cash crops on the hillsides while they plant cacao, achiote (*Bixia oreliana*), chili, tomatoes, and melons on the lower parts (Table 4-1).

Farmers in the valley distinguish between two growing seasons. The first season (*la primera*) begins with the first rains in May and ends in August/September; the second one (*la segunda* or *postrera*) lasts from September/October to December/ January. The dry season lasts from January to April.

Maize is primarily grown on the hillsides and the dry plains in both growing seasons. On the floodplains maize is sometimes planted in the dry season (Table 4-1). To use the legume Mucuna (*Mucuna spec.*) as a cover crop together with maize is a common practice in the valley (see section 4.4). In the *primera* rice is cultivated on the flood plains for sales outside the valley. Beans are usually grown in the second season on the hillsides and the dry plains, either in monoculture or intercropped with maize. Yields vary between 0.9 and 0.5 t/ha, depending on the system (Hernández Jiménez and Silva Goméz 1994). In beans no improved varieties are used while in rice high yielding varieties are planted by more than 85% of the farmers (Hernández Jiménez and Silva Goméz, 1994).

Table 4-1: Farming systems in the Polochic Valley

Growing season	**Hillsides**	**Dry Flat Lands**	**Flood Plains**
Crops grown in the *first season* *(May-Sept.)*	maize, coffee, citrus, cardamom (Mucuna)	maize, chilli, yucca, (beans) horticultural crops (Mucuna)	rice
Crops grown in the *second season* *(Oct-Dec.)*	maize, beans, coffee, citrus, cardamom (Mucuna)	maize, beans, chili, yucca, horticultural crops (Mucuna)	maize, beans
Crops grown in the dry season	(Mucuna)	(Mucuna)	maize, beans

Source: own investigation

The two main growing seasons differ in a few characteristics that lead to some differences in farmers' behavior. Peak of rainfall occurs during the *primera,* which results in a large number of inundated fields on the floodplains and sometimes also on the dry plains. 41% of interviewed farmers have fields in areas threatened by flooding, which means that they need additional fields in the hillsides to cover their food requirements during the first season. This reduction in overall area suitable for cultivating crops increases the pressure on the hillside areas during a time in which they are anyhow especially vulnerable to soil erosion. The higher amount of rainfall in the *primera* also leads to an increased occurrence of diseases making the second growing season more favourable for crop production. Furthermore, prices, especially for maize, are higher in this season giving farmers an additional incentive to produce (L. Chavez, personal communication 1998).

Farmers begin field preparation in April before the first rains. Vegetation on the fields is cut and left for decomposition. Burning of crop residues is still considered a necessary practice by a number of farmers as it is said to reduce pest and weed problems (see Section 4.3.4). Most crops are planted by making a hole with a planting stick into which the seed is dropped. The soil is not moved. Weeding is usually done manually twice during the cropping period. In general smallholders use fertilizer and pesticides only in certain crops (see also Section 4.4). They also harvest all crops manually (Fournier 2000, own investigation 1998).

The majority of small-scale farmers is not able to leave any part of their fields in long-term fallow (more than one growing season). They have to farm each field in at least one cropping season (61% of surveyed farmers). Farmers that have land not used for annual cropping leave it either as a forest or forest/orchard with different fruit trees (50% of surveyed farmers), as long-term fallow (37%) to be used for planting annual crops in a few years or as a natural pasture (13%) (own investigation 1998).

Often smallholders raise chicken, pigs, ducks and turkeys, which are either consumed by the family or sold. All small livestock is kept on the farm. There are only a few smallholders that own cattle, which are then kept near the farm complex (7% of surveyed farmers). As the big cattle producers have fenced pastures, communal grazing, which is typical for many other regions in Central America, is not common in the valley (own investigation 1998).

4.3.4 The erosion problem in the Polochic Valley

Erosion is a visible problem in the hillsides of the Polochic Valley and reports of farmers that their soil is "tired" and yields are declining are often heard. In the survey conducted for this study 72% of the interviewed farmers answered that soil erosion occurred in some parts of their farm. Nevertheless, only 21% considered it to be one of the three most important problems they encountered on their farms (own investigation 1998).

One of the main reasons for the high occurrence of erosion problems in all of Central America is the continuing use of soil degrading farm management practices, like the burning of crop residues in hillside areas or the shortening of fallow periods. Burning is still a common practice in the Polochic Valley, where about one quarter (24%) of the maize-producing farmers interviewed for this study burned their fields in 1997 before the first rains (Table 4-4). They usually burn in the first growing season only as at the beginning of the second season the residues are too wet. The two most important reasons for this practice according to all interviewed farmers are that fewer rodents can be found in burnt fields (29%) and that there will be fewer ear-rot infected plants (21%) (Table 4-2). Furthermore, farmers mentioned that there are less weed problems in burnt fields (15%), fewer problems with insects and other plagues (15%) and sowing is easier (15%). Other reasons include, for example, that the young maize plants are not able to grow well among all the residues and that the maize will grow more equally after burning.

Table 4-2: Main reasons given by farmers in the Polochic Valley for burning crop residues as part of their field preparation

Reasons for burning of crop residues*	% of interviewed farmers
Fewer problems with rodents	29
Fewer ear-rot infected maize plants	21
Fewer weed problems	15
Fewer insect and other plague problems	15
Easier planting	15

* multiple responses possible
Source: own investigation, N = 137

An indicator of how serious soil loss in hillsides can be after burning the crop residues can be found in a study a Swiss student conducted for his Masters Thesis in the valley in 1998. The objective of the study was to investigate the differences in soil loss occurring under various soil management practices (Krebs 2000). In special erosion plots at three different sites with different inclinations he compared the following practices:

1. Cutting of crop residues, burning and planting of maize.
2. Cutting of crop residues, leaving residues as mulch on fields and planting of maize.
3. Cutting of crop residues, leaving residues as mulch on fields, planting of maize and planting of the legume Mucuna 40 days after the maize.
4. Control plot under fallow.

Lost soil and agronomic data concerning maize production were measured over one cropping season. In all three sites soil losses were significantly higher when the plots were burned. On average 1.7 t/ha were lost during the four months in which the trial was conducted. Under other treatments than burning between 0.7 and 0.3 t/ha eroded; differences between treatment 2), 3) and 4) were not significant (Zurek and Krebs, 1999). Nevertheless, there were no significant differences found in parameters concerning soil fertility, like maize yield, plant height, weight of the corncobs, biomass production etc. However, it is hypothesized that differences will be found in these parameters if the trial is conducted over a longer period.

An additional reason for soil degradation problems in the valley is the reduction of long-term fallow periods of six to ten years, which usually characterizes slash-and-burn-systems. Without these long-term fallows the soil fertility cannot recuperate and a decline in yields is the result. Increased demand for arable land in the valley leads many farmers to use their fields continuously and to abandon the system of rotation among their fields (Chavez 1994, O. Garcia, A. Villafuerte and J. Cortéz, personal communication 1998). Thus, for example, in 1997 47% of the farmers had at least one field in which they planted maize in both growing seasons consecutively (Table 4-4) (own investigation 1998).

However, farmers also see important advantages for crop production if they abandon the burning of crop residues. Asked for changes that farmers in the Polochic Valley have noticed since they abandoned burning almost half of the farmers (44%) observed a reduction in soil erosion and 15% reported higher yields (Table 4-3). Other important changes noted were a higher organic matter content of the soil (11%), higher soil humidity (10%) and that the soil gets less hard in the summer months (6%). Nevertheless, some farmers reported that they noticed more pests (4%) and that more time is necessary for planting (2%). One farmer even mentioned that now he observed different weeds growing in his fields.

The majority of farmers, who do not burn anymore in the valley, abandoned burning within the last three years (21%). Some of these farmers though mentioned that they are still burning every second or third year or whenever the amount of residues is very high. Only 10% of farmers have not burned in the last 20 years. The rest stopped with this practice between 4 to 20 years ago.

20 % of the farmers experimented themselves with the No-burning technique. Nevertheless there exists a wide range of further sources of information about this practice: 18 % of the farmers had heard through the public extension service DIGESA about no-burning practices, another 17% from other farmers. Other family members were source of information for 11% of the farmers. The rest of the farmers named extension workers of the NGO CARE International, courses, people from the municipality, foreign aid workers, church organizations or the radio.

Table 4-3: Main observations by farmers in the Polochic Valley concerning changes in their fields after they stopped burning crop residues for field preparation

Observations of changes in non-burnt fields*	% of farmers
Reduction in soil erosion	44
Higher maize yields	15
Higher organic matter content of the soil	11
Increase of soil humidity	10
Soil less hard in summer months	6
More pest problems	4
More time needed for planting	2

* multiple responses possible
Source: own investigation, N = 137

4.4 The Maize and the Maize-Mucuna System in the Polochic Valley

Maize is the main staple of Guatemala and the main crop grown by smallholders in the Polochic Valley. Of the maize producing farmers surveyed in this investigation 91% planted maize in the first and 87% in the second growing season (Table 4-4). Traditionally farmers tried to avoid using the same plot of land to cultivate maize in both cropping seasons and, especially in hillside areas, rotated among their fields. As mentioned earlier, increased pressure on today's arable land drives farmers to abandon this system (O. Garcia, A. Villafuerte and J. Cortéz, personal communication 1998). Thus about half of the surveyed farmers (47%) planted this crop in both seasons in the same field without being able to leave it in fallow. Furthermore, the introduction of a new white maize hybrid, ICTA HB 83, which is very well adapted to the growing conditions of the valley, is thought to increase the incentive for farmers to grow maize for surplus sales on the market.

About one fifth of the maize-producing farmers grow also beans, which are mainly used for home consumption. Rice, grown by 22% of the interviewed farmers on the flood plains during the first cropping season, in return serves as a cash crop, together with coffee and cardamom. As mentioned earlier, cows are seen as a sign of wealth as usually only the big farmers can afford to keep them. Thus only a very small percentage of small-scale farmer owns a cow (7%), which constitute another cash crop for farmers.

Table 4-4: Main farming practices of surveyed maize producing farmers in the Polochic Valley in 1997

Farming practices	% of farmers
Burning of crop residues for soil preparation	24
Planting of maize in *primera*	87
Planting of maize in *segunda*	91
Planting of maize consecutively in the same field in both growing seasons	47
Planting of beans	20
Planting of rice	22
Planting of coffee	8
Planting of cardamom	4
Having land without annual crops (fallow, forest, pasture, etc.)	36
Raising cattle	7

Source: own investigation, N = 137

The number of farmers that do not have to crop all of their farm area annually and can leave parts of their land in fallow, as forest or as pasture has decreased in the last years. In 1997 only 36% still had some land not planted to annual crops.

In Table 4-5 characteristics of maize producing farmers and their farms can be found. With an average farm size of 2.9 ha the surveyed farms can be classified as subsistence farms, who only produce enough to cover the food requirements of the farm family. Even the biggest farmer with about 26 ha can only be put in the group of smaller medium size farmers. This confirms the trend described earlier that big farmers changed their land use from maize production to other uses (mainly cattle ranching). Almost two thirds of the surveyed farmers farm less than 3 ha, while only 1 % of farms are bigger than 10 ha. About one half of surveyed farmers cultivate land in hillside areas, out of which about one third have more than 50% of their farmland in hillside areas. About one fifth of the farmers farm only land in hillside areas. About 40% of the interviewed farmers have land that is inundated during the first cropping season. On average almost 80% of their farmland is inundated in this particular season, obliging them to grow maize in the dry season using the remaining soil humidity. Only about one third of farmers possess some kind of a secure land title, but usually the land title does not cover the whole farm area, but on average merely 77% of it.

Table 4-5: Farm characteristics of surveyed farmers in the Polochic Valley

Characteristics	Level
Average farm size	2.9 ha (4.2 mz*)
Minimum and maximum farm size	0.2 ha (0.3 mz) / 25.9 ha (37.0 mz)
Percentage of farmers with 0.2 ha – 2.9 ha	66 %
Percentage of farmers with 3.0 ha – 9.9 ha	33 %
Percentage of farmers with 10.0 ha – 26.0 ha	1 %
Percentage of farmers with land in hillsides	51 %
Percentage of farms with more than 50% of farm area in hillsides	33 %
Percentage of farms with 100% of farm area in hillsides	18 %
Percentage of farmers that have land in inundated areas	41 %
Average percentage of farm area inundated in *primera*	79 %
Percentage of farmers that possess secure land titles	33 %
Average percentage of farm area with secure land title	77 %

*1 manzana (mz) = 0.7 ha, Source: Own investigation, N = 137

The average age of interviewed farmers is relatively high at 46 years (Table 4-6). Additionally education levels are low, as more than half of the farmers never went to school. Most of the ones that have had any formal education did not even finish primary school.

Households are with on average 6 members quite large. Usually the farm household head works on the fields together with the older male members of the household. Women usually take care of the house and small livestock, but only help with few field activities (weeding and harvest).

About half of the farmers have taken on jobs outside their own farm activities, working on coffee or cattle plantation. This off-farm labor provides an important part of the overall farm income to the households.

Table 4-6: Household characteristics of surveyed farmers in the Polochic Valley

Characteristics	Level
Average farmers age	46 years
Average years of schooling	2.1 years
Percentage of farmers that never went to school	52 %
Average household size	6.4 members
Average family labor endowment	2.4 workers/family
Percentage of farmers with off-farm income	49 %

Source: Own investigation, N = 137

4.4.1 The Maize Cropping System

The area dedicated to maize is significantly smaller (at 5% level) in the first growing season than in the second (Table 4-7), though farmers usually have a plot in the first and in the second season. Reasons for this are a reduction in cropping area due to inundation of the flood plains, increased pest occurrence and lower prices during the *primera.* This results in many farmers growing maize mainly to cover the food requirements of their families, but not for sale within or outside of the valley. If maize is sold it is mainly second season maize. (L. Chavez, personal communication 1997). In the dry season maize is only cultivated by a few farmers on the flood plains using residual soil humidity. The average area of 0.2 ha is very small.

As can be seen from Figure 4-3[1] farmers react to the agro-climatic differences in the two cropping periods as mentioned earlier. Crop residues are burned only in the first season for land preparation as in the second season residues are too wet. Furthermore, a higher percentage of farmers plant hybrids in the *segunda* (first season 27%, second season: 47%). Here, especially the well-adapted

1 Data for the biggest maize plot of each farmer, data for first season taken from Farmer-Subset for *Primera*, for second season from Farmer Subset for *Segunda*, % indicates percent of surveyed farmers (Mucuna users and non-users) using the practices.

Table 4-7: Characteristics of surveyed farmers in the Polochic Valley with respect to their maize cropping system

Farm Characteristics	Level
Average total area planted to maize in 1997	2.1ha (3.0 mz)
Average area planted to maize in *primera* 1997	0.8 ha (1.2 mz)**
Average area planted to maize in *segunda* 1997	1.1ha (1.5 mz)**
Average area planted to maize in dry season 1997	0.2 ha (0.3 mz)

*1 manzana (mz) = 0.7 ha, ** Areas significantly different at 5% level
Source: Own investigation, N = 137

white hybrid ICTA HB83 released in the valley about eight years ago has gained in importance and has led to substantial yield improvements. Since many farmers grow the hybrid mainly for commercial purposes, its use has been associated also with an increase in continuous cropping of fields (see also above). Therefore, it is thought that the use of the hybrid has led to an abandoning of Mucuna planting. In the Maize-Mucuna-System the field has to lie in Mucuna fallow in at least one growing season (for further details see the next section).

The use of yield improving inputs like fertilizer and pesticides is also slightly higher in the second season. These small modifications in cropping practices together with reduced flooding and pest occurrence due to less precipitation in this season lead to higher average yields (first season: 1.6 t/ha, second season: 2,4 t/ha). It can be noticed as well that the hybrid HB83 yields substantially more than the other varieties in both cropping seasons. Nevertheless, for optimal results this high-yielding variety requires the use of fertilizer. This coincides with results of variety trials conducted in the valley over the period of 1996 to 1998 (Fournier 2000).

Nevertheless, it can also be seen that the results show relatively high standard deviations suggesting that the impact of soils, climate and management practices on outputs is quite high.

4.4.2 The Maize-Mucuna System

In the discussion about environmental innovations, productivity enhancing techniques with additional resource conserving characteristics (PERC technologies), and low input agriculture in tropical small-scale agricultural systems, cover crops/green manures have received a lot of attention. Fast growing, herbaceous legumes with high aboveground biomass production are usually selected and are

Figure 4-3: Maize Cropping System in the Polochic Valley in the first and the second cropping season

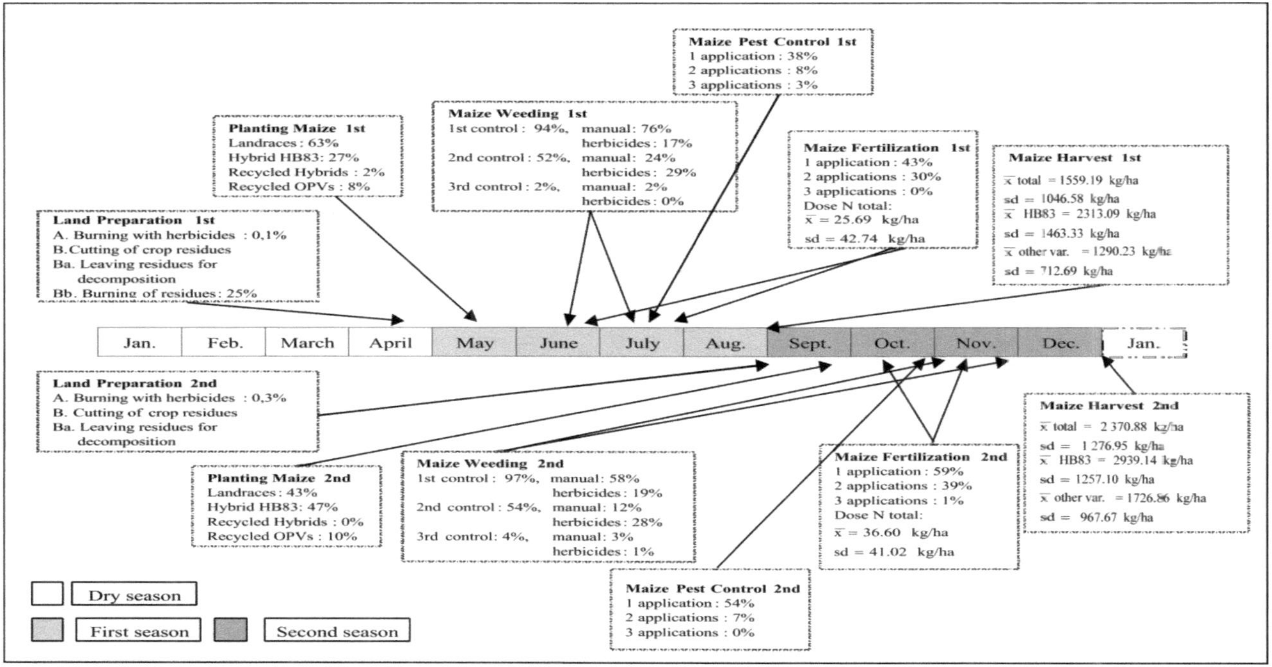

Data for the biggest maize plot of each farmer; % indicates number of surveyed farmers in each data subset using the practice

Source: own investigation, N = 137

either intercropped or planted as a relay crop or in sequence with the main crop. The main crop can be either a staple or a plantation crop. The aboveground biomass is usually slashed and left on the ground as mulch. Parts of it can also be utilized as fodder or food. By providing a thorough soil cover relatively fast soil erosion is effectively reduced. Additionally, weeds are suppressed, nutrients supplied to the main crop and soil moisture conserved. Nevertheless, the species and their planting systems have to be selected carefully so that they do not compete with the main crop (Eilittä 1998).

Cover crops/green manure systems became an interest of researchers in Mesoamerica in the 1980's and 1990's. Different species were researched for their potential use together with maize, the main staple in the region (Zea *et al.* 1991, Barreto 1994). The legume Mucuna (*Mucuna spp*) was of special interest as it was already known in the region since about the 1920's. It had been spontaneously adopted by farmers in parts of Guatemala, southern Mexico, and Honduras for it's beneficial yield effects in maize. That is why farmers in the region sometimes call it "Frijol de Abono" = "Fertilizer bean" (Eilittä 1998, Buckles *et al.* 1998). Agronomic research confirmed these promising results for maize production in the humid subtropical and tropical zone of the region (Gordón *et al.* 1993). Today Mucuna use is promoted by a large number of organizations in all of Central America. Adoption factors influencing Mucuna use have been investigated for the northern part of Honduras (Sain and Buckles 1998, Buckles *et al.* 1998). For Guatemala, from where its use probably had spread into Honduras and Mexico (Buckles *et al.* 1998), it was of interest to study these factors with a special focus on the use of Mucuna as a PERC technology. In the following section an overview of this plant and its use in the Polochic Valley is given.

4.4.2.1 Description and History of Mucuna

Mucuna (*Mucuna spp*) is a climbing, annual legume, whose English and Spanish name "Velvetbean, Frijol Terciopelo" originates from the pubescence of its pods that gives it a velvety feeling when touched. A variety of different species exist, whose exact distinctions and taxonomy are still unclear. The main differences between the species include the color of the seeds (black, white, beige, gray and mottled, in the Polochic Valley white and black seeds are used), the thickness of the pubescence and the number of days until seed maturity (between 100 to 300 days). Mucuna grows best under humid, warm conditions, below 1500m of sea level. It can tolerate a variety of abiotic stresses, like low soil fertility and soil acidity, but is sensitive to frost (Buckles *et al.* 1998). In addition, Mucuna seems to be photoperiodic as it reacts to day length in flowering and seed production. For Honduras and also Guatemala it has been reported that flowering starts with shorter days in October, independently of the planting dates. After seed production the plant dies naturally, 45 – 60 days after flowering (Buckles *et al.* 1998, L. Chavez, personal communication 1998). The seeds contain the toxic compound L-Dopa, halucinogenic typtamines, phenols and tannins (Ravindran & Ravindran 1988, Awang et al. 1997, all in: Buckles *et al.* 1998), which can be neutralized through cooking and washing of the seeds before consumption (Eilittä 1998). Mucuna is known for its vigorous growth of vines (up to several meters) and foliage, which can produce between 5 to 12 t/ha of dry-matter and makes it well suited as a cover and forage crop (Triomphe 1996).

Mucuna originated in southern China and eastern India, where it was cultivated as a vegetable for its edible seeds. It was probably used in other parts of Asia also as a soil improvement practice and forage crop and reached the USA in the 1870's, where it was widely adopted in the Cotton Belt in maize-cotton rotation system. From there it was first introduced into Central America in the 1920's through its use on the banana plantations as fodder for the mules of the United Fruit Company, who owned extensive tracts of land along the Atlantic coast of Guatemala and Honduras. Here the Kek'chi Mayans, who worked in the plantations, became familiar with this legume (Buckles *et al.* 1998, Eilittä 1998).

Mucuna was first introduced into the Polochic Valley by a Jamaican farmer growing banana in the area in the 1930's (C. Chavez, personal communication 1997, Buckles *et al.* 1998). Since then farmers in the valley have used Mucuna intercropped with second season maize; large-scale adoption though only seemed to have occurred in the 1950's (Buckles 1995 in Triomphe 1996). This coincides with a larger influx of Ladino farmers into the valley, who established farms mainly on the fertile flood and dry plains, pushing smaller farmers up the hillsides (C. Chavez, personal communication 1997). Thus a similar development described for the northern part of Honduras, where small-scale farmers had to give way to large cattle farms and where Mucuna has been utilized extensively during the last two decades, could have taken place in the Polochic Valley. For Honduras it was hypothesized that smallholders were looking for alternative ways of producing maize under marginal, hillside conditions and thus began to value the productivity enhancing effects observed when Mucuna was present in the field. Through observation and trial-and-error they might have learned how to manage best this new production system under their local conditions. Its use then was spread by word of mouth, observation and migration among the farmers of the area (Triomphe 1996). Today Mucuna is widely known to farmers in the Polochic Valley though it is mainly used by small-scale maize producing farmers. Nevertheless, its utilization by the subsistence maize producers seems to have declined in the last decade, going hand in hand with a decline in fallow periods and increased continuous cropping of fields (Chavez 1994, O. Garcia, A. Villafuerte and J. Cortéz, personal communication 1998).

4.4.2.2 Benefits and disadvantages of Mucuna use

The utilization of Mucuna has various beneficial effects on soil properties and soil fertility, which can be summarized as follows. An in-depth investigation of these issues for the Litoral Atlantico region of Honduras, which has similar agro-climatic conditions as the Polochic Valley, was conducted by Triomphe 1996. Most of the results presented here are taken from his investigation, if not stated otherwise.

- As all legumes Mucuna lives in a symbiotic relationship with *Rhizobium* bacteria, which have the ability to fix the nitrogen from the air and then supply it to the plant in exchange for other nutrients. Thus Mucuna biomass can contain under favorable growing conditions up to 250 kg/ha of nitrogen, which will be released from the decaying biomass between 35 to 80 days after the plants die or are cut down (Lathwell 1990 cited in: Eilittä 1998, Gordón *et al.* 1993, Chavez 1995). It is estimated that at least about half of this amount is taken up by the maize crop and by weeds. Depending on whether Mucuna is left as a mulch or incorporated the amount of N-fertilizer supplied to the main crop has been estimated in experiments under varying conditions to be between 20 to 150 kg N/ha (Lobo Burle *et al.* 1992, Van Noordwijk *et al.* 1995, Carsky *et al.* 1998, all cited in: Eilittä 1998). Additionally, other plant nutrients like potassium and phosphorus are supplied to the maize in quantities sufficient to cover its requirements.
- Physical soil properties, like structure, water infiltration capacity, organic matter contents, and the biological activity in the soil can also be improved by the use of this legume.
- The thick layer of biomass produced helps to conserve soil humidity and to suppress weeds. Additionally, the soil is protected from the erosive impact of the heavy rains, basically stopping soil erosion in the fields.
- Farmers do not burn their fields as they usually lie under a Mucuna cover during the first cropping season in which fields are burnt for soil preparation (Figure 4-5). Additionally, the Mucuna layer is relatively easy to slash.
- Mucuna also seems to possess nematicide and allelopathic qualities helping in the suppression of nematodes and weeds.

All the described qualities of Mucuna lead to a stabilization and an increase in maize yields (up to 100% reported in field experiments in various places), thus reducing the risk of crop loss for the farmer (Triomphe 1996). Nevertheless, this beneficial effect on yield depends on avoiding competition between Mucuna and the main crop since it has been reported that maize yields can be affected adversely if the Mucuna is planted too soon after maize planting (Zea *et al.* 1991). In the Polochic Valley though farmers seem to have developed a system in which this competition is avoided.

The Maize-Mucuna-System also allows the continuous cropping of fields without long-term fallow periods usually necessary to maintain soil fertility especially under hillside conditions. As Triomphe (1996) reports from his investigations in the Litoral Atlantico of Honduras, where he evaluated the long-term changes in soil properties under more than 11 years of continuous Mucuna use, maize yields on "old" Mucuna fields were as high or higher than yields on "young" fields. Compared with non-Mucuna fields yields doubled. He concludes that from an agro-ecological point of view the Mucuna system seems to be a viable alternative for farmers seeking sustainable ways of maize production under marginal hillside conditions, in which maintaining soil fertility becomes an important problem if long-term fallow periods have to be shortened due to pressure on arable land.

The Mucuna-System possesses further favorable characteristics for farmers in Central America. The system is relatively easy to manage and can be quite easily integrated into the cropping practices of maize production in the region. Henceforth the farmer does not have to learn too much additional know-how to understand and implement the system. In general, there is relatively little additional labor necessary, especially if the system is well established and Mucuna re-seeds itself. Also no additional cash resources have to be used to maintain the system and establishment costs are very low as well, as most farmers are able to obtain Mucuna seeds from members of their communities, if needed. Labor requirements for land preparation are said to be lower in the Mucuna system as the slashing of the Mucuna layer is easier and faster than that of other natural fallow systems (Buckles *et al.* 1998).

Despite of all the favorable qualities of Mucuna, Mucuna can also bring some difficulties for the farmer. One of the problems arises from the long development period of the legume until seed maturity. In Mucuna fields maize can only be grown during one growing season while Mucuna occupies the field during the rest of the year (Figure 4-5). Thus in intensive production systems, where farmers want to use their field for maize or other crops twice a year, the opportunity costs of land are relatively high (Triomphe 1996). Additionally, even in a less intensive system, the farmer needs at least one other plot of land to grow maize for home consumption during the season in which the field lies in Mucuna fallow, if this field is not big enough to produce sufficient supplies during the one growing season.

Another disadvantage of the Maize-Mucuna-System is the two to three year time span of growing Mucuna until its beneficial effects on yields and soil properties can be seen (Almendares *et al.* 1995). Thus farmers have to invest in the system at a net cost during this time (Sain 1998). This time period is definitely shorter than that of many other conservation technologies but nevertheless it might be too long for farmers with a shorter planning horizon, e.g. landless farmers that have to rent land in each growing season. Thus the incentive for these farmers to use Mucuna is not very big, unless the other farmers in the area are utilizing it too. Here is a good example that it is important to investigate the circumstances and the cropping system with which the Maize-Mucuna-System is compared to determine the potential benefits for its users.

Furthermore, there have been a few reports of landslides on very steep hillsides under dense Mucuna cover after very heavy rains. Reason for this probably is the very dense root system close to the soil surface but the lack of deeper roots that could hold the heavy soil-Mucuna plate under these extreme conditions (Almendares *et al.* 1995, Triomphe 1996). Nevertheless, the occurrence of this kind of landslides was never reported for the Polochic Valley.

4.4.2.3 The Maize-Mucuna System in the Polochic Valley

Mucuna use in the Polochic Valley: In 1997, 48% of the farmers (66 farmer) in the Polochic Valley interviewed for this study planted Mucuna in at least one of their fields (Figure 4-4). 21% of the interviewed farmers (43 farmer) had used it in the last ten years but had abandoned its use in 1997. And about one third of the farmers (28 farmer) never used it.

To avoid competition with young maize plants Mucuna is sown in-between the maize rows about 45 to 70 days after maize planting and then left to grow during the dry and the following cropping season. The varieties used in the valley (black or white seeds) need about eight to nine months until seed develop-

Figure 4-4: Mucuna use in the Polochic Valley in 1997

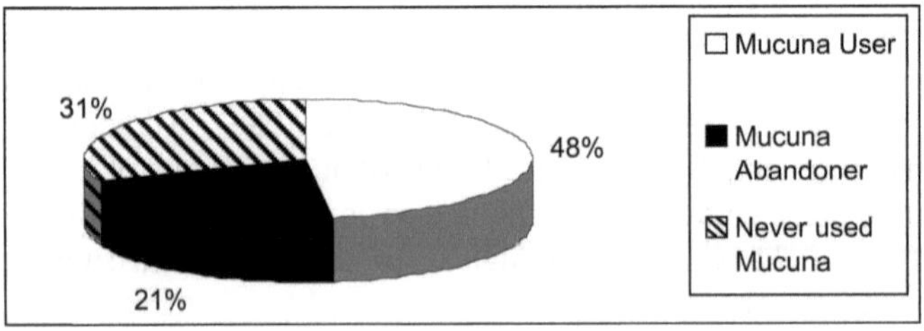

Source: own investigation, N = 137

ment. Farmers cut the vines before sowing maize in the next year. The residues are left on the ground, seeds are collected if needed. Sometimes the legume reseeds itself. Maize is planted directly into the decaying biomass, which supplies nitrogen and other nutrients to the new crop. In this system the farmer is able to use his field only once a year while it lies under Mucuna-fallow the rest of the time.

Due to the development cycle of Mucuna, farmers in the Polochic Valley intercrop it with the maize of the *Segunda* (Figure 4-5). In this case it is sown in December/January in-between the maize rows, which were planted in September/October. Mucuna continues growing until it is cut in September/October of the following year after flowering and seed production, when preparing the field for the next second season maize. Nitrogen is released from the decaying biomass at the right time for uptake by the new maize. By subsequently suppressing weeds and conserving soil humidity Mucuna can help to increase maize yields up to 55%, as results of ICTA research in the valley conducted in 1992/1993 demonstrated. But the field can only be used once a year for maize production (Chavez 1994, Chavez 1995).

If Mucuna is planted together with first season maize (Mucuna sowing in June), Mucuna flowers and sets seeds in October/November due to the photoperiodic sensitivity mentioned earlier. As the plants die afterwards, nitrogen is released at the beginning of the dry season in January. This then does not result in a significant increase in yields as the next maize crop is only planted in the following April/May (Chavez 1994, Chavez 1995). But Mucuna still helps in this way to reduce soil erosion as it provides a protective cover for the soil. Thus the 'traditional system' of intercropping Mucuna is with second season maize. Nevertheless, farmers today also experiment with first season planting. In this case they either leave the field under Mucuna fallow during the second season and the following dry season or cut the Mucuna vines after harvesting the first season maize and plant maize or some other crop into the Mucuna residues. In this manner they can use their field twice a year. The beneficial effects of Mucuna though are not fully achieved for the reasons mentioned before.

Given the history of Mucuna diffusion in the Polochic Valley it is not surprising that the majority of farmers (62% of surveyed farmers), who plant the legume today (1997), use it already for more than 10 years in their main maize field (Figure 4-6).

Figure 4-5: The Maize and the Maize-Mucuna System in the Polochic Valley

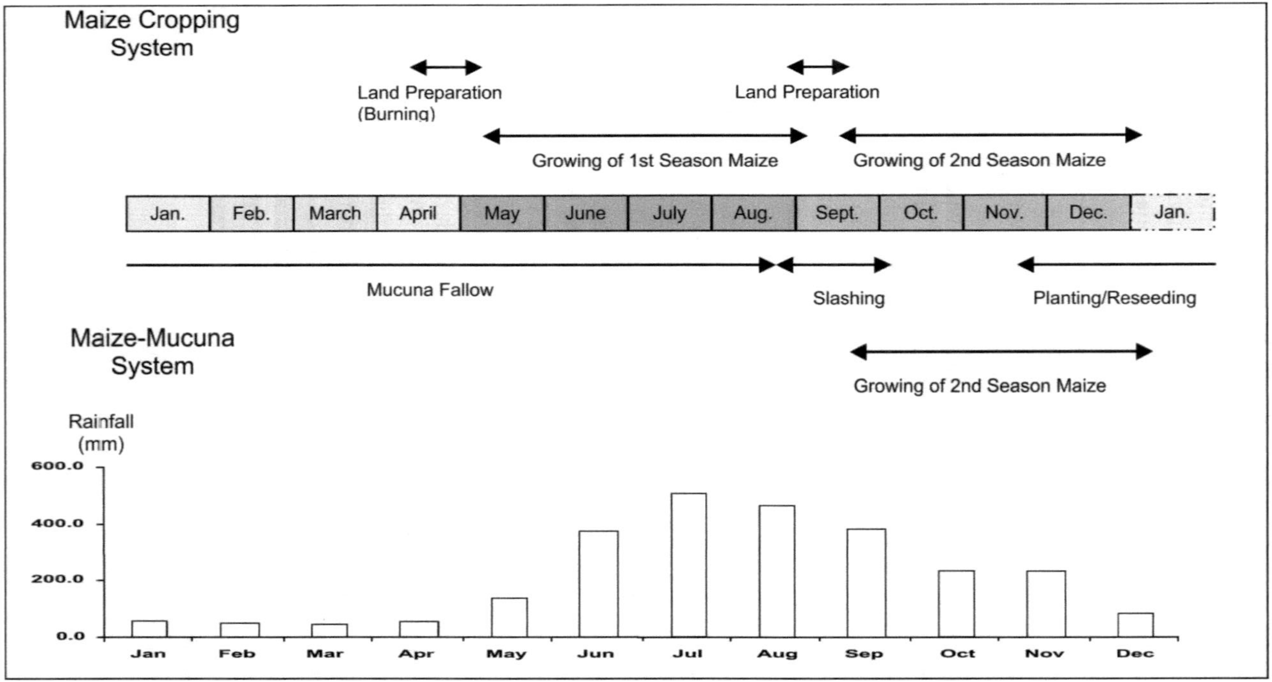

Source: Own investigation

After a gap in adoption this practice seems to have become attractive again for farmers as about one third of the surveyed farmers only adopted the legume in the last four years before the investigation in 1998.

Figure 4-6: Years of Mucuna use in the Polochic Valley in 1997

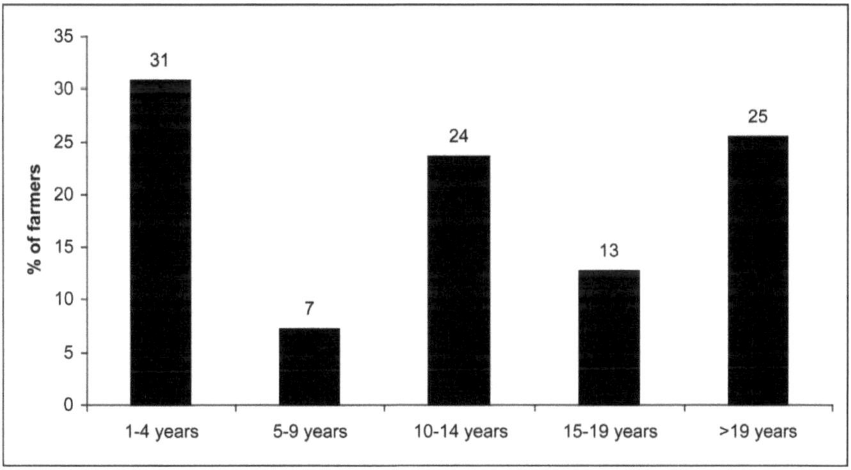

Source: own investigation, n= 66

Differences between Mucuna users, non-users, and abandoner: As can be seen in Table 4-8, for Mucuna users this legume is foremost a good fertilizer (79% of users) and 36% of the surveyed users associate it directly with higher yields of the maize crop. Additionally, farmers value the reduction in weeding time (29%) as the fast growth of the legume provides strong competition for most weeds. Nevertheless only very few Mucuna users (4%) also see the beneficial effects of Mucuna as an erosion control measure. Therefore, farmers value the legume mainly as a productivity enhancing technology, while its resource conserving aspect is almost overlooked. For most scientists though Mucuna is foremost an erosion control measure and is promoted as such.

Farmers mainly give three reasons for not planting Mucuna (Table 4-8): Either they cannot get the seeds (26% of non-users), Mucuna is not suited for their maize field (e.g. due to inundation in the first growing season (25%)) or the farmer needs the field to grow another crop in the season in which his field would otherwise lie under Mucuna-fallow (23%). It is interesting to note that almost all farmers are convinced of the beneficial effects Mucuna planting would

Table 4-8: Reasons given by surveyed farmers for their use, abandonment or non-use of Mucuna

Reasons FOR Mucuna use (n = 66)	**% of farmers***
Good fertilizer	79
Higher yields	36
Less weeds	29
Less erosion	4
Soil is softer	4
Other reasons	6
Reasons AGAINST Mucuna use (n = 43)	**% of farmers***
No seed available	26
Mucuna can not grow in the field for ecological reasons (inundations etc.)	25
Farmer needs field for other crops	23
Missing management information	9
More work	7
Farmer grows maize on rented land	5
Seed to expensive	5
Mucuna does not have any benefits	2
Other reasons	4
Reasons for ABANDONING Mucuna (n = 28)	**% of farmers***
Farmer needs field for other crops	36
Mucuna can not grow in the field for ecological reasons (inundations etc.)	29
No seed available	21
More work	4
Mucuna does not have any benefits	4
Higher yields can be achieved with chemical fertilizer	4
Farmer grows maize on rented field	4

* multiple responses possible
Source: own investigation, N = 137

have for their fields, as only 2 % of non-users think that Mucuna would not contribute to yield. And only a very small number of farmers do not want to plant Mucuna because their land is rented (5%). In addition, the majority of non-users do not seem to think that this technique implies a large amount of additional work or that the seeds are too expensive. Therefore, the non-use of the legume seems to have more to do with structural factors like seed availability and information on the technology than with characteristics of the technology itself or access to land. The only exception is the fact that due to increased pressure on arable land farmers are no longer able to have a real Mucuna fallow. This becomes even clearer when we look at the reasons for farmers to abandon the use of the legume. The three reasons given by non-users are here still named as the most important ones as well, but their order has changed and the increased intensification of land use has gained in importance (36% of abandoners). Again increased labor requirements or the fact that farmers grow maize on rented land is of minor importance. Here we can see that the Polochic Valley is undergoing overall structural changes that also influence the maize production system. The tendency is going, like in many other parts of the region, towards a higher intensity of land use and a shortening of improved fallow periods.

There exist a number of differences between Mucuna users, non-users, and abandoners. One important point is their use of improved maize seed material, in particular the use of the Hybrid HB 83 and its effects on the Maize-Mucuna-System. As explained earlier the use of the hybrid is hypothesized to be associated with more continuous cropping of fields, which results in an intensification of the farming system, and the abandonment of the legume planting. Figure 4-7 shows that the use of maize landraces is still quite widespread in the valley, since up to half of the interviewed farmers use them in at least one of their maize fields. Nevertheless, it can be seen that Mucuna users are actually the group with the lowest percentage with land planted to these varieties (48%), while they are using the hybrid HB 83 as much as for example, the farmers who never planted the legume. They also use quite a few other improved hybrids or Open Pollinated Varieties released earlier in the valley. This signifies that Mucuna users are the group that uses improved varieties the most. As we have seen already, for Mucuna users Mucuna is mainly valued for its fertilizer effect. Therefore, Mucuna is a cheap possibility to replace nitrogen fertilizer, necessary for the good growth of improved varieties. The hypothesis that the use of improved varieties is directly linked to abandoning the Mucuna system is thus not correct. For the calculations of the Logit Model though, explained in the next section, the hypothesis is maintained to get a better understanding of the changes occurring in the valley.

To add a bit more information on the differences between Mucuna users, non-users and abandoners, which might be important for building the analytical models, the knowledge of the three groups on soil conservation techniques is investigated. Mucuna users are the group that is the most likely to use other kinds of soil conservation techniques apart from Mucuna (37%) (Figure 4-8). These include mainly live barriers with Madrecacao (*Gliricidia sepium*) and contour

Figure 4-7: The use of different maize varieties by Mucuna users, abandoners and non-users in the Polochic Valley in 1997

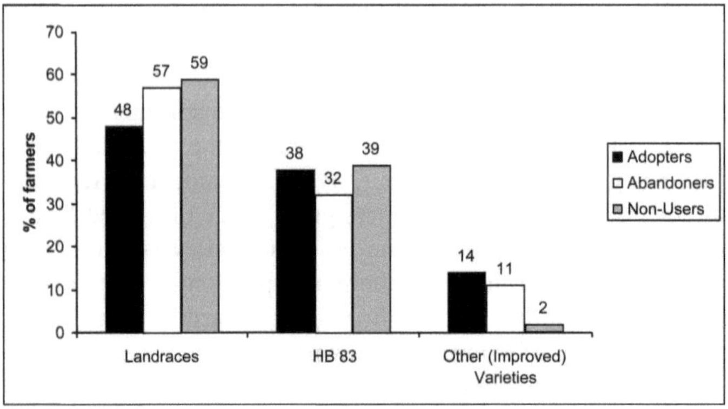

Source: own investigation, N = 137

tillage. Mucuna non-users are the ones that know the least about and use the least of soil conservation practices. But they are the ones that most frequently burn their crop residues before the first rains in May (40%). Also quite a number of Mucuna abandoners (32%) still burn their residues every year, but also quite a few of them have started to introduce measures to alleviate the erosion problem in their fields (20%). As the fields of most Mucuna users lie under Mucuna fallow during the first growing season, it is understandable that the vast majority of them do not burn their fields any more, as this would also threaten the beneficial effects of the technique.

Concerning differences in cropping practices of the three groups (Table 4-9), Mucuna users have on average the largest farm size (4.4 mz = 3.1 ha). Mucuna abandoners have the smallest farms (3.1 mz = 2.1 ha), while farmers that never used the legume are in a middle position (3.7 mz = 2.6 ha). Mucuna users have in both growing seasons on average a higher percentage of the land in hillsides on which they grow maize than the two other groups. This is likely to be one reason why they value Mucuna as it helps to reduce erosion. Despite the fact a higher number of Mucuna users use improved varieties, which in general require fertilizer for good performance, they use less N-fertilizer than Mucuna non-users or abandoners. The fertilizer use is in all groups higher in the second growing season as then more maize is grown for surplus sales. Despite the fact though Mucuna users apply less fertilizer and grow more maize in the less favorable hillside areas, their maize yields are in both growing season as high as the ones of abandoners and non-users. This indicates the positive effect that Mucuna has

Figure 4-8: The use of soil conservation practices other than Mucuna and the use of crop residue burning techniques by Mucuna users, abandoners and non-users in the Polochic Valley in 1997

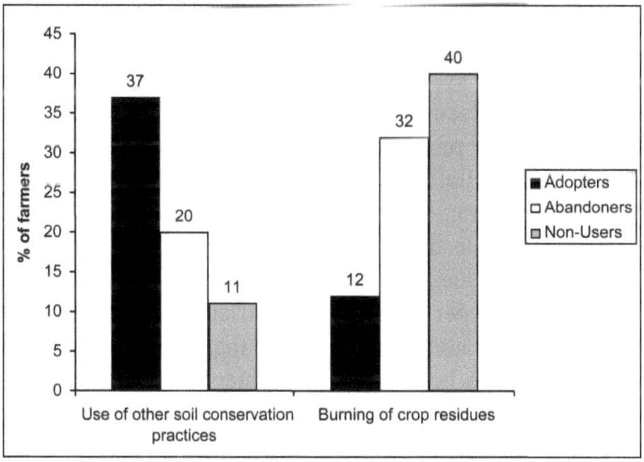

Source: own investigation, N = 137

Table 4-9: Characteristics of the maize cropping system of Mucuna users, abandoners, and non- users

Characteristic	Mucuna user (n = 66)	Abandoner (n = 28)	Non-User (n = 43)
Average farm size	3.1 ha (4.4 mz*)	2.1 ha (3.1 mz)	2.6 ha (3.7 mz)
Average % of maize fields in hillsides in *primera*	42	22	39
Average % of maize fields in hillsides in *Segunda*	25	13	17
Use of improved maize varieties	52%	43%	41%
N-fertilizer use in *primera*	21.2 kg/ha	27.5 kg/ha	35.1 kg/ha
N-fertilizer use in *segunda*	36.1 kg/ha	48.8 kg/ha	39.8 kg/ha
Average maize yield in *primera*	1.5 t/ha	1.8 t/ha	1.3 t/ha
Average maize yield in *Segunda*	2.3 t/ha	2.4 t/ha	2.3 t/ha

* mz = manzana (1 mz = 0.7 ha)
Source: own investigation, N = 137

on maize output and goes along with the reasons given for Mucuna planting. Farmers are able to achieve the same output with less input costs.

In summary it can be said that the Mucuna system was and is quite beneficial for farmers and was well adapted to the maize production system as long as land was abundant in the area. Therefore it spread in the Polochic Valley on a farmer-to-farmer basis. Today though it seems that parts of the maize production system need to change and a different kind of system needs to emerge for farmers with less land. Nevertheless, this modified system should have the same agronomic and economic characteristics as the Mucuna system in terms of erosion control, fertilizer effects, little additional labor requirements, but fallow periods have to be shorter to allow the use of fields in both growing periods if necessary without jeopardizing soil fertility in the long-run.

In the following we will investigate the characteristics of Mucuna adopters and non-users and the factors that influence their decision with respect to Mucuna use in further detail to gain a deeper insight in the differences between the groups. This is helpful for drawing conclusions about the characteristics of PERC technologies and how they can be adapted to farmers' conditions. In a first step a Logit Model will be constructed to better understand the probability of farmers with certain characteristics to use the legume. This model though does not seem to capture many of the underlying factors that explain the use of this PERC technology. Therefore a Structural Equation Model (SEM) with latent variables is built, in which the underlying factors that influence the farmer's decision to use measures to increase soil productivity in a shorter term (e.g. through fertilizer) versus longer-term techniques, like Mucuna, is modeled.

4.5 Modeling farmers' decision to use Mucuna

The decision of a farmer to plant Mucuna is influenced by a number of different factors that include the biophysical conditions of the farm, certain characteristics of the farmer and the farm household, and the institutional settings under which the farm operates. In addition, the perception of the farmer on the benefits and drawbacks of the technique are important in this respect. Understanding the importance of the different factors can help to explain the farmers' behavior and to develop ideas on how to enhance Mucuna adoption.

As mentioned in section 4.2, in total 137 farmers in the county of Panzós were interviewed on their maize production system, and their use of Mucuna. For understanding the maize system a fairly detailed questionnaire was necessary. Therefore, the sample was divided into farmers who were asked about their maize planting techniques in particular in the first growing season (63 farmers) and those who were questioned in detail about the second growing season (74 farmers) (for the questionnaires see Annex A). Nevertheless, all farmers were asked if and how they use Mucuna and other soil conservation techniques. Interviewed farmers in each sub-sample mainly utilized the legume together with second season maize, but there were also quite a number of farmers that experimented with first season planting. The main analysis is run for the farmers asked about the *segunda*. The models were also run with the detailed data on the first season to see if differences between the two sub-samples could be detected.

4.5.1 The Logit model

The decision of the farmer to plant Mucuna can be seen as a binary choice between the two alternatives of use or non-use. Which strategy is followed depends on different attributes of the farmer and his farm. Knowing about the significance of different characteristics in the farmer's choice can help us to estimate the probability that a certain farmer in our target population will use this practice. From an econometric point of view, the dependent variable, which represents the choice of the farmer, is in this case not continuous but represents a discrete 0 (= non-use)/1 (= use) choice. The expected value of the dependent variable can be interpreted as the probability that a particular farmer with certain characteristics will plant Mucuna. (Kennedy 1992). This probability though can only take values between 0 and 1.

Estimating the probability of adoption according to different farmer's characteristics can be done in various ways. A Linear Probability Model is one option, which estimates in a linear regression the dependent, discrete variable Y, depending on certain farmers' characteristics X (see equation 1). Results for Y from the estimation can be translated into the adoption probabilities (Maddala 1983).

$$Y_i = \beta x_i + u_i \quad \text{with } E(u_i) = 0 \text{ and } x_i = \text{farmer's characteristic} \qquad (1)$$

Nevertheless, there are a few drawbacks to this model, which at one time was preferred by many researchers for its computational ease. The estimated probabilities might fall outside the range of 0 to 1, which results in wrong estimates for the adoption probability. In addition, the error term u_i is not normally distributed and may have heteroscedasticity problems. This can then result in inefficient estimators of β, when using the OLS estimation method (Maddala 1983, Kennedy 1992). Therefore, today mainly non-linear estimation techniques are used. Probit and Logit models are here the prevalent choices, as with them estimated probabilities fall between 0 and 1. A Probit model is based on a cumulative *normal* probability function, while a Logit model uses a cumulative *logistic* probability function (Pindyck and Rubinfeld 1981). Both of these sigmoid curves run between 0 and 1. Both models lead to similar results, as only the scale of the β coefficients is different and the logistic probability function has a fatter tail (Pindyck and Rubinfeld 1981, Maddala 1983, Davidson and MacKinnon 1993). As the Logit model is easier to calculate (Pindyck and Rubinfeld 1981), it is often preferred to the Probit and is also used in this study to calculate Mucuna adoption probabilities, subject to characteristics of farmers in the Polochic Valley.

The Logit model can be specified as follows (equation 2) (Pindyck and Rubinfeld 1981):

$$P_i = F(\alpha + \beta X_i) = 1 / 1 + e^{-(\alpha + \beta X_i)} \qquad (2)$$

with e:natural logarithm (= 2.718),
P_i: adoption probability, given certain farmer's attributes
X_i: attribute of i-th farmer

This equation can be transformed into equation 3 (Pindyck and Rubinfeld 1981, Kennedy 1992), which depicts the logarithm of the probability that a farmer chooses one of the two alternatives:

$$\text{Log } P_i / 1 - P_i = \alpha + \beta X_i \qquad (3)$$

The Logit model can be estimated for two different kinds of data: grouped or ungrouped data (Kennedy 1992). In case a grouping is possible and the sample is large enough OLS estimation can be used. In case a grouping is difficult or the sample is not big enough, one can also employ maximum likelihood procedures, which are today readily available in most statistical packages (Pindyck and Rubinfeld 1981). In this study a maximum likelihood procedure is used to estimate the following equation (equation 4); for this the statistical program STATISTICA was used:

$$\text{Log } (P(\text{use}) / 1 - P(\text{use})) = ß_0 + ß_1 X_1 + ß_2 X_2 + \ldots + ß_n X_n \qquad (4)$$

with X_1 to X_n = different farm household characteristics

Hypotheses: Under the circumstances of small-scale maize producing farmers in the Polochic Valley, the following hypotheses were developed on farmer's characteristics X_1 to X_n influencing the adoption behavior (see also Table 4-10):

Mucuna has a long history in the valley and a large number of farmers have used it. Therefore older farmers are thought to have long-term experience with the legume, knowing about its benefits and disadvantages. Hence it is hypothesized that they are more likely to plant the legume in their fields.

Table 4-10: Variables, expected signs and hypotheses for the Logit Model of Mucuna adoption in the Polochic Valley

Variable	Expected sign	Hypothesis
Age of farmer	+	Older farmers are more likely to use Mucuna as they have more experience with this system that is already know in the area for a long time.
Family labor endowment	-	As the Maize-Mucuna system is a labor saving technology, a farmer with little family labor to rely on is more likely to use it than one with a large family labor endowment.
% maize fields with slope	+	The higher the % of maize fields in hillsides the more inclined is the farmer to use Mucuna as he is more likely to face erosion problems and unstable yields than farmers in the flatter zones.
% inundated fields	-	The higher the % of fields inundated during one season, the less likely is the farmer to use Mucuna as it can not grow there and he needs the rest of his land for food production.
% of land with secure land title	+	Farmers, who have secure land use rights, are more likely to use Mucuna as they usually have a longer planning horizon.
Farm size	+	The bigger the farm, the more likely is the farmer to use Mucuna as he has more land available to take into production while the rest lies under Mucuna fallow in one season.
% of land without annual crops	+	The higher the % of land without annual crops, the more likely is the farmer to use Mucuna as he has more land available to take into production while the rest lies under Mucuna fallow in one season.
Use of white maize hybrid HB 83	-	The Maize-Mucuna system seems to be replaced by an intensified production system, in which farmers plant maize twice a year using the hybrid HB83. Thus farmers growing the hybrid are less likely to use Mucuna.

Source: Own investigation

The aggressive growth of Mucuna biomass can suppress weeds quite successfully and therefore reduces the weeding time for farmers in the maize crop substantially (Triomphe 1996, Sain and Buckles 1998). If farm households face a labor shortage due to low family labor endowments they will look for ways to decrease the workload. Therefore they will value a reduction in weeding time due to Mucuna planting.

Fields in hillside areas are more prone to soil erosion than in the flatter parts of the valley. Hence soil fertility problems are more likely to occur in hillsides. Since Mucuna is a good erosion control measure and additionally boosts soil fertility (Triomphe 1996, Eilittä 1998), it is hypothesized that the higher the percentage of hillside fields, the more likely a farmer is to plant Mucuna.

Inundation of fields in the flood plains, which reduces the size of arable land, is a serious problem in the first growing season. Therefore farmers that have a high percentage of their land in inundation zones are less likely to use Mucuna as it will not be able to grow there. In addition, farmers need their fields in other agro-ecological zones to produce sufficient food. This reduces the likelihood that they will leave these fields under Mucuna fallow in the *primera.*

There has been a lot of discussion about land titling and its effect on resource conservation (see for example Baland and Platteau 1996, Vosti and Reardon 1997). One general notion is here that if farmers have clearly assigned land titles it will increase their incentive to conserve soil fertility and the resource base of the land they own. This will also lead to a longer planning horizon of farmers, as they know that they will be able to use their land for longer than one cropping season. Therefore it is hypothesized for the Polochic Valley that the higher the percentage of fields with a secure land title, the more likely a farmer is to plant Mucuna.

Since Mucuna is a land-intensive technique (Sain and Buckles 1998, Buckles et al. 1998), farmers need a certain amount of land to leave under Mucuna fallow in one of the growing seasons, while they can produce food on the other fields. Thus the bigger the farm, the more inclined farmers are to use the legume system as they can rotate among their fields. Additionally, if the farmer has a part of his farm under non-annual crops he might as well be more inclined to use the legume as he can always take some of this land into maize production while another part lies under Mucuna fallow.

As explained earlier, the hypothesis that the use of the hybrid HB 83 is leading farmers to abandon the Maize-Mucuna-System is maintained for the Logit model analysis. Here it was thought that since hybrid users might be more inclined to intensify their cropping system and plant maize in both cropping seasons, making it impossible for them to plant Mucuna as well.

Results: Table 4-11 shows the results of the Logit Model for the data sub-set of farmers interviewed mainly about their maize production system in the second growing season, which is the main Mucuna planting season.

The χ^2 value, which is significant at the 1.0% level, and the p-level of 0.001 suggest a good model fit. In addition, this is confirmed by the percentage of 0s and 1s that are predicted correctly.

Table 4-11: Results of the Logit Model for the data-subset of farmers interviewed in detail about the SECOND growing season

Variable	Coefficient	Standard Error	p-Level
Constant	-2.65*	1.54	0.09
% Inundated area	-2.80*	0.91	0.00
Farmer's age	0.04*	0.02	0.09
Use of hybrid HB83	1.32*	0.67	0.05
% Maize in hillsides	0.15	1.23	0.90
Family labor endowment	0.23	0.22	0.30
% Land with secure title	0.37	0.79	0.64
Farm size	-0.06	0.14	0.68
% Land without annuals	-0.12	1.85	0.95

n = 74, χ^2 = 25,79**, df = 8, p-Level = 0.001, % 0s predicted correctly = 59.7, % 1s predicted correctly = 40.3, * significant at 5 % level
Source: Own investigation

The percentage of inundated fields in the first growing season is a significant factor influencing farmers' behavior not to grow the legume. The higher the percentage of inundated fields the more likely it is that the farmer needs the other fields for food production in this season and can not leave them in Mucuna fallow (negative sign as expected). The farmer's age is also a significant attribute to influence Mucuna use and as expected the older the farmer is, the more likely he is to plant the legume. A third significant factor to lead farmers to grow Mucuna is the use of the maize hybrid HB 83, which is positively correlated with Mucuna planting. This is at first surprising as it had been hypothesized that the more intensive hybrid production system is replacing the Mucuna system. But already when looking at the differences between Mucuna users and non-users and abandoners it became clear that Mucuna users are the ones that use the highest percentage of improved seed material. Additionally they also have on average a bit more land than the other groups, which makes it easier for them not to plant all fields continuously. Therefore it seems that Mucuna users know about the fertilizer requirements of improved seeds and have found in Mucuna a relatively cheap way of fulfilling these requirements, at least partly.

As expected, the sign for the variable'% of maize in hillsides' is positive, leading to the conclusion that farmers with hillside maize fields are more inclined to use Mucuna. Nevertheless, the variable is not significant and can therefore not be seen as an important factor influencing farmers' behavior. The number of family members helping with weeding and other farm chores as well does not seem to influence the farmer's behavior, despite the fact that farmers gave as one reason for planting Mucuna that the fields are less weed infested. Furthermore, the sign was unexpectedly positive. Also the percentage of land with clear property rights does not seem to be significant for the farmer's decision. The positive sign though indicates that farmers are more likely to use the legume if they have secure access to land. The availability of land does not seem to influence the farmers decision very much either. Both the variables 'Farm size' and '% of land without annual crops' are not significant. And different than expected, the signs are negative, meaning the smaller the farm or the percentage under fallow, forest etc. the more likely the farmer is to use the legume, despite the Mucuna system being a land-intensive technique. The reason for the negative signs is therefore difficult to understand.

The Logit Model was also run for the data sub-set of farmers who had been interviewed in more detail about their maize production system in the first growing season. Detailed questions about their overall use of different crops were also posed. Therefore, for almost all variables data were available for the overall cropping system. Table 4-12 shows the results for these farmers.

The model fit for this dataset is also quite good as the X^2 distribution here is also significant at the 1% level. The overall p-level for the model is with 0.006 as well satisfying. Nevertheless, the overall fit of the model was a bit better for the dataset of farmers interviewed in detail about the second season.

Different from the second season data sub-sample, for this set of farmers a few other reasons seem to be significantly influencing their decision to use Mucuna. Here the percentage of maize fields in hillside areas is a significant variable, which is, as expected, positively correlated with the planting of the legume. Also the amount of fields with a secure land title and the farm size are important factors influencing the farmers' decision. In both cases the signs are positive, meaning that the higher the amount of land available to the farmer and the more secure his access rights to the land the more likely he is to use the legume. This goes along with the postulated hypotheses. The only variable, which is significant and also has the same sign in both sub-samples, is 'Farmer's age'. So the older a farmer is more likely to use Mucuna.

Table 4-12: Results of the Logit Model for the data subset of farmers interviewed in detail about the FIRST growing season

Variable	Coefficient	Standard Error	p-Level
Constant	-5.36*	1.80	0.00
% Maize in hillsides	2.15*	1.33	0.11
% Land with secure title	1.82*	0.96	0.06
Farm size	0.31*	0.21	0.14
Farmer's age	0.07*	0.03	0.01
Family labor endowment	-0.01	0.27	0.97
% Land without annuals	2.04	1.79	0.26
% Inundated area	0.80	1.20	0.51
Use of hybrid HB83	-0.63	0.76	0.41

n = 63, χ^2 = 21,34**, df = 8, p-Level = 0.006, % 0s predicted correctly = 44.4,
% 1s predicted correctly = 55.5, * significant at 5 % level
Source: Own investigation

Non-significant variables for this farmer sub-sample are both 'Family labor endowment' and the '% of land under non-annual crops'. In the both cases the signs though are correct. For family labor endowment this means that the less family labor there is available to the farmer the more likely he is to plant Mucuna. In the case of land without annual crops, the farmer is more likely to plant the legume if he has a lot of land under fallow, forest, etc. For this data sub-sample the area of inundated fields in the first growing season is also not significant. In addition, the positive sign of this variable is confusing here. Another non-significant variable is the one modeling the use of the maize hybrid HB 83. Additionally the sign here is negative, suggesting that farmers planting the hybrid are less likely to use the legume, which goes along with the original hypothesis. In this sub-sample though farmers were asked about their use of the hybrid in the first growing season. As explained earlier for this cropping season the use of the hybrid is not as high as for the second season, due to pest problems etc. Additionally Mucuna fields would lie under fallow in the first season, suggesting that farmers that want to plant maize in this season are less likely to plant the legume at the same time, though there are a few farmers that now experiment with intercropping Mucuna with first season maize.

Both model runs revealed a number of factors significantly influencing farmers' use of Mucuna. Farmer's age is here definitely one of the decisive variables, but certain attributes of the maize cropping system, like the use of hybrid seeds or the percentage of hillside or inundated fields, also play an important role in farmers' decision. In addition, a number of structural factors like farm size and land titling can become influential under certain circumstances. All these factors therefore play a role in the farmers' decision to use longer or short-term soil improvement practices and influence the adoption and continuous use of Mucuna.

Nevertheless, the results of the Logit model are not yet satisfying for explaining the farmers' decision to use a PERC technology like the Maize-Mucuna-System. The model does not capture well enough the complex factors underlying the decision to use rather short- (e.g. fertilizer) versus long-term (Mucuna) soil fertility improvement measures. No really decisive factors were identified. Some of the interactions between factors such as farm circumstances and the decision to use a practice aimed at maintaining soil productive potential in the future cannot be modeled with enough clarity. Therefore, a Structural Equation Model with latent variables was constructed, which is particularly aimed at revealing underlying soil management decision factors and interactions between these factors.

4.5.2 The Structural Equation Model with latent variables

Structural Equation Modeling (SEM) can be described as a powerful multivariate analysis method, which is composed of a set of different analytical techniques. These include certain parts of path analysis or causal modeling, confirmatory factor analysis, multiple regression analysis and correlation and covariance structure analysis. The basic idea of SEM is that hypothesized causal, linear linkages between a number of variables can be tested and the weights of the variables on each other estimated by analyzing their variance and covariance (STATISTICA Electronic Manual 2002).

SEM has its roots in Path Analysis developed in the 1920s by the geneticist Sewell Wright (Loehlin 1998, Maruyama 1998). Wright built a unidirectional causality chain to model the impact of each set of parental genes on their offspring. To verify the hypothesized links he set up simultaneous equations to describe the correlations between the variables. In the 1960s the ideas of Path Analysis were taken up in social science research, mainly in behavioral and psychological research, which looked into the relationships between different social variables and social status, education levels or job choice. Here still mainly unidirectional causalities between the variables were modeled using again simultaneous equations or matrix algebra for the estimation process. Only with the development of better computers and computer programs (e.g. LISREL or AMOS) did it become possible to estimate more complex models that also incorporated feedback loops between variables (non-recursive models). For these models not just ordinary multiple regression techniques were utilized, but also so-called indirect least square, two- and three stage least squares estimation methods emerged. These models were increasingly combined with ideas on how to incorporate unobservable (latent) variables into the estimation process to explain certain observed behaviors or choices made (Maruyama 1998). Today SEM is mainly used in psychology, sociology, political sciences, education (Schumacker and Lomax 1996), and marketing research.

In this brief explanation of the history of Structural Equation Modeling with latent variables we find all the elements employed today in this approach. The underlying idea of SEM is that we can build hypotheses on causal relationships between certain exogenous and endogenous variables. A change in exogenous variable A causes a change in endogenous variable B. To verify the causal chains a linear regression model can be utilized in which the magnitude of the change is estimated. The relationships can be described using a path diagram, which depicts the different variables and their hypothesized linkages. Usually researchers want to understand the relationships between a whole set of variables that can be linked by direct or indirect relationships and even a number of feedback loops. For these a number of structural equations can be developed to model these simultaneously. Scientists can then not only look at a number of exogenous variables and see how well they explain the observed variance of the endogenous variables but also understand the respective weight of each predictor variable (Maruyama 1998). From there another step can be taken in case we have a number of variables that are not directly observable. These so-called latent variables, which for example describe the reasons for observed behaviors or decisions, can only be depicted by a number of observable indicator variables. Like in factor analysis they together are an expression of an underlying factor or variable. Used in SEM we therefore have a number of indicator variables describing an unobservable latent variable. Between the latent variables causal relationships can be hypothesized which are then tested by estimating how much a change in an exogenous latent variable influences an alteration in an endogenous latent one. Thus it can be differentiated between an external (indicator) and internal (latent) information structure (Nuppenau and Hedden-Dunkhorst 1998), which helps to illustrate the relationships of interest and the weights of different factors. SEM can therefore be a powerful analytical tool whenever perceptions or opinions play an important role in the described system, as this method provides a possibility of modeling these and their outcomes. In the following we will look at the different parts of SEM in more detail.

4.5.2.1 Theoretical background

Causality analysis: Causality analysis has been used mainly by social scientists to investigate causal linkages between different factors that explain a certain type of behavior. Its purpose is to understand the reasoning behind and the reasons for a certain decision. Thus factors that are sometimes not directly visible or connections between certain factors that on first sight do not seem to be linked can be made visible. In this way causality analysis can help decision-makers to assess the outcome of a certain decision or to develop policies and strategies. Or scientists might be able to develop solutions for a problem by understanding the links and driving forces behind it.

To begin the analysis the researcher has to establish a theoretical model, which contains the relevant factors necessary to explain the problem he/she is confronted with. Then linkages between these factors have to be described. Thus each model consists of a number of endogenous, dependent variables that are influenced by different exogenous, independent variables (both types of variables have their corresponding error terms). To establish a causal relationship between two variables X and Y three conditions have to be fulfilled: 1) There has to exist a covariance between variable X and Y. 2) There has to exist a temporal sequence, also called temporal asymmetry, between X and Y, which means that one has to occur before the other one. 3) The covariance between X and Y has to hold also if the variables earlier in the causality chain that influence both of them, so called "confounding variables", are removed. This means that other possible causal factors that might have caused the observation of an existing link between X and Y have to be eliminated without that the linkage breaks down. This of course might not be easy and requires certain simplifications (Asher 1976, Schumacker and Lomax 1996). Each causal relationship can be described mathematically by structural equations, which usually do not have a constant term if standardized data are used. The relationships can also be shown graphically in a causality diagram (see Figure 4-9).

To illustrate the way a causality model can be constructed the following example will be used (Figure 4-9): Suppose a researcher wants to understand the factors and their possible links that influence the decision of a farmer to adopt a certain technology. He/She has identified three main factors that seem to play a pivotal role in the farmer's decision: Membership in a farmers organization (A), education level/years of formal schooling (B), age of farmer (C). Factors A and B are hypothesized to directly impact on the decision while factor C only has an indirect influence through B. A and C are exogenous variables as they are not affected by any other variable and influence themselves B and the variable "Farmers Adoption Decision", which are called endogenous variables. The R_i's re-

Figure 4-9: Example of a causality model for farmers' technology adoption (modified after Asher 1976)

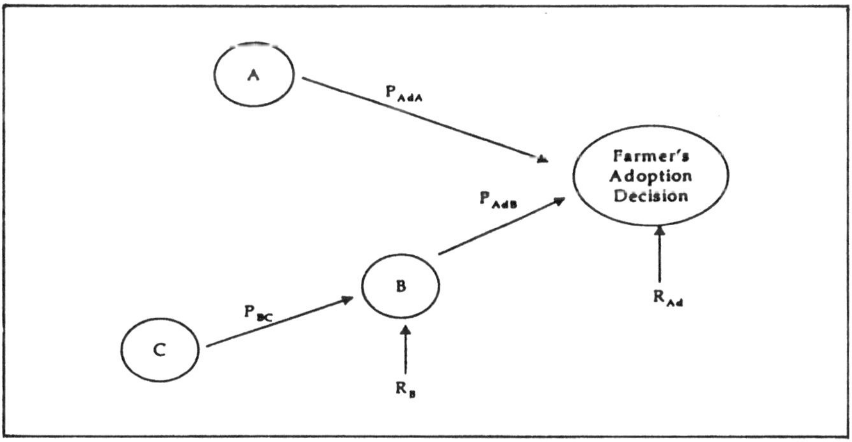

present the error or residual terms of each endogenous variable, which contain factors that were not measured but still might impact on these variables (see also condition for causality No. 3) (Asher 1976). In our example, these factors could be for variable B if the farmer had to work on the farm as a child and was thus not able to go to school all the time. For the variable "Farmer's adoption decision" this could be the case if the farmer cultivates more than a certain amount of land and if he is eligible to receive credit at the local bank. If with further investigation it becomes evident that the access to credit of the farmer impacts very strongly on the adoption decision this factor has to be included as an additional exogenous variable in the model. The P_i's in Figure 4-9 above the arrows depict the weight that one variable has on another one, thus measuring the influence of this variable. Obtaining robust values for these weights is very important as they actually show the underlying causal relationships the researcher is looking for. Understanding the linkages between the variables could then lead to the development of strategies or solutions for the investigated question. For our example this would mean the following: A strong relationship between the adoption of the investigated technology and the membership in a farmers' organization has been calculated (because, for example, the farmers' organization provides the labor necessary to implement the new technology through a system of mutual labor exchange). Contrary to this the influence of the education level of the farmer on his decision is rather low (because the technology is relatively simple and does not require the understanding of very complex biological processes). Thus an institution that would like to promote the investigated technology in an area should seek the support of existing farmers' organizations or encourage the forming of such groups. Measures alone to enhance the understanding of the technology (like courses etc.) would not be sufficient. From the above mentioned it becomes clear that causality modeling can help to understand the causal processes determining observed behaviors and to specify the impact of certain factors, which then can be used to develop solutions for the investigated problem (Asher 1976).

Path Analysis and Structural Equations: A powerful method to calculate the weight of one variable on another, which is what the researcher is interested in, is to employ path analysis– an estimation technique to assess not only the existence of the hypothesized causal relationships between the variables but also their strength. This procedure therefore does not help the researcher to develop the theoretical model but it estimates the strength of the specified theoretical relationships and therefore verifies the theory (Schumacker and Lomax 1996). If this method is used the weights, or P_i's in Figure 4-9, are called "path-coefficients" and can be seen as partial regression coefficients (Schumacker and Lomax 1996, Maruyama 1998). For their estimation multiple regression analysis is employed. Each endogenous variable is regressed on each other variable that is thought to have a direct influence on this specific variable (Asher 1976, Maruyama 1998). For the path-coefficients tests of significance can be calculated equally to other regression methods (e.g. t-values). Additionally, the goodness-of-fit of the specified model can be tested. Here the discrepancy between the observed and the calculated correlation matrices is computed and tested for its significance using χ^2 and other statistics (see below). A significant value for χ^2 means that the observed and the calculated correlation matrices do not fit each other, leading to the result that the theoretical relationships specified in the path model do not fit the observed data (Schumaker and Lomax 1996). In addition, there are a number of other goodness-of-fit criteria that need to be employed. They will be explained in further detail in the following.

In our example the path-coefficient P_{AdA} would be calculated by running a regression between variable A and the variable "Farmer's adoption decision". The same can also be done for any endogenous variable and its corresponding residual term R_i to assess its impact on the variable. By calculating the path coefficient for R_i with standardized data the variation in the particular variable can be obtained that is not explained by the factors depicted in the model. This is due to the fact that, if standardized data are used, the variance of the variable equals 1. The residual path coefficient is defined as $(P_I)^2 = (1 - R_i^2)^2$ with R_i^2 as the multiple regression coefficient that usually represents the explained part of the variance of the variable. Thus $1 - R_i^2$ is the unexplained portion of it and the residual path coefficient is the square root of this expression (Asher 1976). With path analysis also the indirect effects of one variable upon another can be estimated. In the above example it is hypothesized that the age of the farmer only has an indirect impact on his adoption decision through his education level. This could be the case if older farmers had less access to schooling and thus received fewer years of formal education. This in turn makes it more difficult for them to understand the requirements and the processes behind the functioning of the technology. How important the age of the farmer is for his level of education and with this also for his decision-making can be seen by the magnitude of the path coefficient P_{BC}.

The correlation between two variables can also be decomposed into its simple and compound parts; indirect effects are always compound paths within the specified models. A compound path equals the product of the simple paths of which it consists. For the decomposition a set of rules, called “Wright’s rules”, exist which specify how paths have to run from one variable to another: 1) The path to an endogenous variable can only pass the same variable once. 2) The path cannot go against the direction of an arrow if it first went forward (in the direction of an arrow) on a different arrow. 3) The path cannot go through an unanalyzed arrow between two exogenous variables (double-headed arrow) more than once (Loehlin 1998).

Structural equations can be seen as an alternative way of expressing the causal linkages (Loehlin 1998). Instead of in a path diagram they are written as a set of equations in which each variable at the end of an arrow is a function of all the variables from which arrows point to this variable. For our example (Figure 4-8) the following structural equations can be written:

$$B = P_{BC} * C + R_B$$

$$\text{Farmer's Adoption Decision} = P_{AdA}*A + P_{AdB}*B + R_{Ad}$$

Latent Variables: Often in socio-economic research the factors thought to be influential in certain decisions and behaviors are not directly observable or measurable; they are hypothetical constructs or so-called “latent variables” (Schumacker and Lomax 1996, Maruyama 1998). Nevertheless, factor analytical procedures can be used to evaluate the possibility that a number of measurable indicator variables share “common variance-covariance characteristics” (Schumacker and Lomax 1996, p.45) and therefore define an underlying, latent factor. These indicators or manifest variables then act as exogenous variables that influence one endogenous (latent) variable. This approach is also used in SEM, as here usually most of the variables of interest are latent ones. Hence the researcher establishes a so-called ‘measurement model’ (Schumacker and Lomax 1996), which indicates which indicator variables identify which latent variable.

Then, using confirmatory factor analysis, this theoretical model and the hypothesized interactions can be tested. If the hypothesized relationships cannot be verified in this approach other indicator variables have to be tested or the latent variable has to be redefined completely. If we return to our example on the adoption decision of a farmer to adopt a new technology we might be able to rewrite the example with latent variables. Let's suppose that factor A in Figure 4-8 now represents perception of the farmer on the food requirements of his family, factor B is his attitude towards risk and factor C represents the wealth of the farm family. All these variables are theoretical constructs that embody concepts, opinions and perceptions that are not directly measurable. But all can be seen as underlying factors that impact on the farmer's decision. Each of the latent variables can be assigned a number of indicators, which all together define the variable. For factor A this can be, for example, the family size (supposing that larger families need more food than smaller ones), if the farmer is able to sell surpluses of staple crops he grows (this implies that the farm produces sufficient food for the family) and if the farm has any animals that have to be fed by farm products as well. In a diagram latent variables are usually depicted in a circle while the indicators are shown in a rectangle (see also Figure 4-10).

Structural Equation Modeling with latent variables: In the SEM approach with latent variables all the different parts described above come together. A path diagram is used to depict the hypothesized causal relationships between the latent factors. They can be expressed as structural equations to be estimated for calculating the strength of the relationships. This part of the model is then the structural model. Additionally the diagram shows the indicator variables, which identify the different latent variables. This part is called the measurement model. Figure 4-10 gives an example of an SEM. As can be seen, different from other econometric models that explain one dependent (endogenous) variable with a number of independent (exogenous) variables this model tries to establish links between a whole set of exogenous variables and various endogenous ones between which exist a set of specified relationships (Nuppenau and Hedden-Dunkhorst 1998). The indicator variables (x's and y's) constitute a vector for each latent variable (ζ's and η), which are then used to calculate the covariance matrices for the relationships between the latent factors. Therefore four different types of variables can be differentiated in this kind of model: Exogenous latent and observable ones and endogenous latent and observable ones (Table 4-13). Each of these variables has its corresponding error term. Latent variables are usually depicted by an oval while indicator variables are shown in a rectangle. The interest of the researcher lies in discovering the strength and the significance of the theoretically established relationships by estimating the path coefficients Γ_{11} and Γ_{21}.

Figure 4-10: Structural Equation Model with latent variables (modified after Nuppenau and Hedden-Dunkhorst 1998)

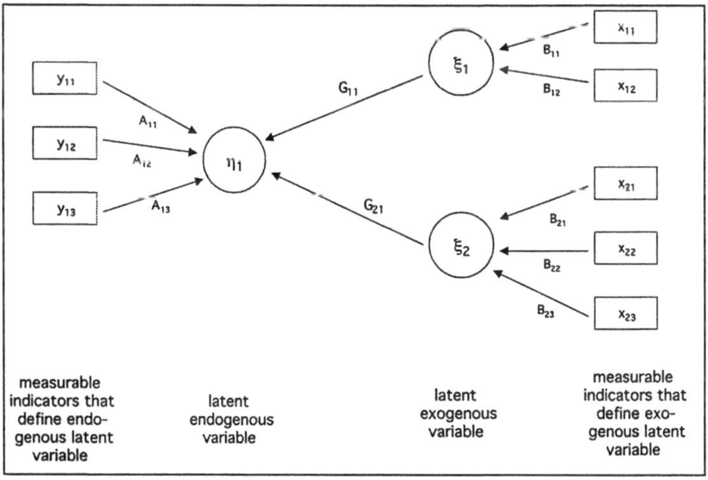

Table 4-13: Types of variables in a Structural Equation Model with latent variables

Type of variable	**Characteristics**	**Function**	**In Diagram**
Exogenous latent variable	Non-observable, theoretical construct	Determines latent endogenous variable(s), not explained in the model	ζ, in circle
Exogenous indicator (manifest) variable	Observable, measurable	Defines latent exogenous variable	x, in rectangle
Endogenous latent variable	Non-observable, theoretical construct	Determined by exogenous latent variable(s), explained in the model	η, in circle
Endogenous indicator (manifest) variable	Observable, measurable	Defines latent endogenous variable	y, in rectangle

Source: Own investigation

Jöreskog (1993) points out that there are different approaches to the use of SEM results: model confirmation, testing alternative models, and generating new models. The original approach is to confirm the theoretically established model, using different measures of Goodness-of-fit to assess the overall model fit. This information can then be used for planning further steps in the research process. As there sometimes exist different possible model specifications that comply with the overall theory framework, testing alternative models can be useful. Nevertheless, this needs to be done with solid theoretical justifications. This is even more important if not just alternative but new models are generated. It is suggested that great care is taken when modifying or changing the model completely as this also implies changes in the developed hypotheses (Maruyama 1998).

Loehlin (1998) and Maruyama (1998) suggest different ways of testing the validity of and modifying the developed SEM. The basic step is to assess the overall model fit. In addition, a model can be cross validated by splitting the data set and running the analysis with different data sub-sets, if the sample size is big enough. When SEMs are used strictly for model confirmation conclusions of the results can be drawn here. If the model does not fit the data, it can be investigated if the problems arise from the measurement or the structural model. To test the measurement model a confirmatory factor analysis can be run. This analysis will clarify if some of the indicator variables used to define the latent variables are causing the problem. Is this the case, it needs to be decided if some of the indicators should be modified. If the misfit is not found in this way, the problem of the model lies in the structural model. This implies that the hypotheses and with this the theory which led to the developed model have to be revised. As mentioned above, this should only be done with great caution and theoretical justification.

The overall fit of the estimated model can be evaluated according to different Goodness-of-fit criteria. Different from other estimation techniques though so far none of these indices has been established as the most reliable one (Schumacker and Lomax 1996). All compare the covariance matrix estimated by a particular model with the covariance matrix of the observed data and their accurateness varies, depending for example on sample size (Schumacker and Lomax 1996, Loehlin 1998). The better the model covariance matrix fits the covariance matrix of observed data the better is the model fit. It is recommended to use a number of indices to assess the overall model fit (Schumacker and Lomax 1996). Maruyama (1996) distinguishes between absolute indices that describe the level of unexplained variance and relative indices, which compare the fit of the specified model with other possible models. The SEPATH module in STATISTICA provides a large number of different Goodness-of-fit indices, from which in this study a few absolute and relative ones were selected.

4.5.2.2 Hypotheses

As we have seen in the Logit Model there are a number of different factors related to the farmer's cropping system on the one side and to certain structural factors on the other side that influence the farmer's decision to use Mucuna. But how do these factors fit into the overall picture of the farmer's decision to use measures aimed at shorter or more long-term maintenance of soil fertility? What are other determinants, which may not be directly observable as the factors mentioned earlier?

According to the Induced Innovation Theory (Chapter 3) the choice of the farmer depends on his resource endowments, the scarcities he perceives and the technological possibilities available to him. Investigating the links between these different factors can therefore shed some light on how these factors play together in the decision of the farmer to use a new technology. His choice of farming practices is as well influenced by his perception of the benefits of a certain technology. This is even more important when determining if and which conservation techniques to implement on his farm as their benefits are often not directly visible, but only after a certain period of time. Here short-term versus long-term soil management considerations are important. Modeling the decision framework of farmers with respect to their technology choices needs to include some of these not directly observable, latent variables. Understanding the connection between these factors will then help in the formulation of policies designed to strengthen or weaken certain links. This can also lead to revealing some of the relationships between the overall policy environment and the decisions taken at the individual level, which is important in the context of the discussion on social benefits versus private costs for sustainable resource management.

Figure 4-11 shows a SEM with latent variables modeling the decision framework of farmers in the Polochic Valley for using long- or short–term measures to maintain soil fertility. In the model four endogenous and five exogenous latent variables are specified with their corresponding indicator variables. In the following this theoretical model and the rationale for choosing the depicted linkages will be described. In a second step the model will be estimated to test the hypothesized causal relationships.

Farmers in the Polochic Valley have different technological option when having to decide about how to boost and maintain soil fertility levels. Depending on the farmer's time-horizon, his intention can be to maintain soil productive potential over a long period of time and to employ certain environmental innovations/PERC technologies, which aim at maintaining functioning agro-ecological relationships (Chapter 3). As soil erosion is a major problem in the area, they include the 'Use of Mucuna in the last 10 years', the No-burning of fields', which is considered to be an easy to implement soil conservation practice, and the 'Use of conservation technologies' available in the valley. In addition, the growing of 'Maize in rotation' with other crops and also the overall 'Number of crops' grown by the farmer can be used as indicator variables to define the latent variable 'Long-term investment in soil fertility'.

If a short-term boost of soil fertility is the farmers' intention, he will use a different set of practices, which belong to the category of commercial innovations (CIs). These include the application of nitrogen fertilizer ('Kg N-fertilizer used') and pesticides ('No. of pesticide applications') and using a high-yielding maize variety (HB 83) ('Use of HB 83'). These indicator variables thus define the latent variable 'Short-term management of soil fertility'.

The implementation of these practices is aimed at reaching a particular production goal. Ruthenberg (1980, p. 3) characterizes a farm as "an organized decision-making unit in which crop and livestock production is carried out with the purpose of satisfying the farmer's goals". In subsistence production systems, such as the maize production system in the Polochic Valley, the primary goal of farmers is to produce sufficient food to guarantee the survival of the farm family. Any surplus production is sold or traded to obtain raw materials and cash income needed to buy farm inputs, clothes, etc., or to pay school fees, taxes, etc. Additionally, many subsistence farmers also sell their labor to bigger farms to increase their cash income. Wealth accumulation can be seen as an additional long-term goal of many subsistence farmers. Therefore, the farmer will use certain farming practices and additional measures to satisfy the food and cash requirements of the household. Depending on how well the farmer is able to achieve this goal and what kind of techniques are available to him he will adjust farming techniques, implement perceived beneficial new practices or take on off-farm labor opportunities. Thus in the SEM the latent variable 'Perception on Satisfaction of Basic Needs' was specified. This variable reflects the level to which the farmer feels that the basic needs of his family, and with this his most important farming goal, is covered. In the context of the maize farmers in the Polochic Valley different indicator variables can serve as observable measures of this satisfaction level. On the one side the 'Family size' indicates the potential demand for food and cash. Whether this demand can be satisfied depends partly on the 'Maize yield' obtained as maize is the most important staple in the family's diet and therefore also the most important crop grown. If and how much the farmer can sell parts of the obtained maize yield is an additional indicator of the satisfaction level, captured in the indicator 'Maize sales'. Furthermore, the cash income level of the farm household can shed some light on the overall satisfaction of basic needs. In the valley most farms obtain additional income through the sales of farm products, captured in the variable 'Income from sales', or by working off the farm, symbolized by 'Off-farm labor'.

The latent variable 'Perception on Satisfaction of Basic Needs' is an endogenous variable, which is influenced by the overall 'Wealth level of the farm household'. The higher the wealth level of the farm family the easier it is for them to cover their basic requirements. Furthermore, the level of 'Human Capital' of the farmer, which is an expression of the years of 'schooling' and the farmer's age, can influence how precarious the farmer views the overall satisfaction level. The production levels achieved by the farmer are of course another important factor influencing the overall degree of satisfaction. This production level is partly determined by the practices employed to boost crop output in the current season and therefore to improve short-term soil fertility. The choice the farmer makes to use these practices hence influences the fulfillment of basic food requirements. On the other side this choice is in itself determined by the perception on the satisfaction level, as the farmer might adjust practices accordingly. Therefore a feedback loop exists here. The 'Perception on satisfaction of basic needs' additionally influences the 'Short-term management of soil fertility' indirectly by impacting on another latent variable called 'Intention of technology use'. This variable is central and captures the farmers reasoning behind his choice of more productivity enhancing (CIs) or resource conservation (EIs) techniques. Visible indicators of this variable are on the one side the 'Reasons why farmers use Mucuna' as a representative of a practice that maintains and enhances soil fertility in the long run. On the other hand the use of the maize hybrid HB 83 ('Advantages of HB 83') can be seen as a technique that is associated with the increase of short-term productivity. Therefore reasons why farmers are using one or the other technique can give us some background on their thinking behind their technology choice.

The variable 'Intention of technology use' is furthermore influenced by a number of exogenous latent variables that reflect characteristics of the farmer, the physical attributes of the farm and its institutional setting. 'Household wealth' determines how pressing the need is to use all available land for food production. Only with a certain wealth level, which guarantees that sufficient food can be produced or purchased for household consumption, the farmer is able to leave parts of his fields in fallow or experiment with new practices that might not produce immediately the same output level as the old techniques. 'Human Capital' also impacts on the farmer's choice of practices by influencing his understanding of new techniques and how they modify the functioning of his production system. The 'Knowledge on conservation issues' is another attribute

Figure 4-11: Structural Equation Model with latent variables modeling the decision of farmers in the Polochic Valley to use short- or long-term soil fertility management measures

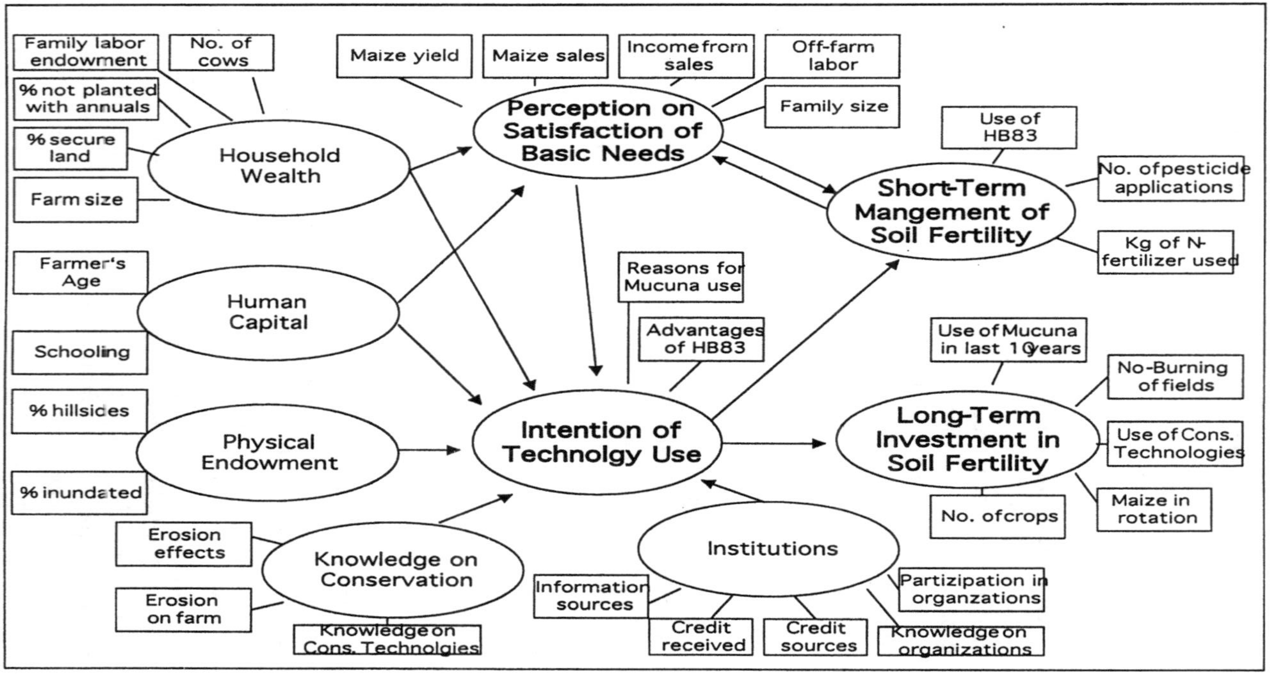

Source: own investigation

of the farmer, which determines the choice of technologies aimed at either short- or long-term increase in soil fertility. Here the occurrence of 'Erosion on farm' and the knowledge on 'Erosion effects' are two decisive factors determining this latent variable. In addition of course the 'Knowledge of conservation technologies' available in the area, their benefits and disadvantages and the costs associated with their implementation and maintenance are of influence here. Another important variable is the 'Physical Endowment' of the farm. As we have already seen in the Logit analysis, in the context of the Polochic Valley technology choice in this respect is mainly determined by the amount of farmland in 'Hillsides' a farmer has or how much of his land will be 'Inundated' during the first cropping season. These indicator variables determine the availability of land and the degree of the erosion problem the farmer faces.

The institutional setting also impacts on the farmer's intention behind his technology choice. Therefore the latent variable 'Institutions' is constituted of the 'Information sources' the farmer uses, his knowledge on 'Credit sources' and his success in receiving credit ('Credit received'). Depending on the wealth level of the farmer he has to use credits for implementing new practices or to buy inputs. For many small-scale farmers in the Polochic Valley though it is very difficult to obtain credit from formal sources due to problems with collaterals. Additional indicators for the variable 'Institutions' are the farmer's 'Knowledge on organizations' and 'Participation' in any of them. Different organizations, such as farmers groups, water committees or church groups, provide the farmer with the possibility to exchange information and learn from experiences of others.

The variable 'Intention of technology use' is shaped by a number of different factors, but it then determines the inclination of the farmer to use more 'Short-term management of soil fertility' measures or make a more 'Long-term investment in soil fertility'. If in the model estimation these linkages turn out to be critical they give a good indication where policy measures to enhance the use of PERC technologies, such as the Maize-Mucuna-System, could be helpful. Policies could then work on strengthening these linkages and the factors, which are significantly shaping the intention behind farmers' technology choice.

4.5.2.3 Results

In this study the confirmatory approach to the use of SEM results was chosen. Therefore, first the theoretical model for the decision-making framework of small-scale maize producing farmers in the Polochic Valley was developed. Then the model was calculated with the main data subset of the farmers interviewed about the second growing season. The results are presented in the following. The model was also run with the data subset of the farmers surveyed for their first season cropping practices as many of them are also using Mucuna in either of the two growing seasons.

In both data sub-sets the data were standardized before running the analysis. Following Schumacker and Lomax (1996), the calculated path coefficients are in this way specific to the particular data set, but they allow a comparison of results within the sample. In addition, missing data were replaced using the sample mean. This was done as otherwise the missing cases would have been deleted, which would have reduced the small data set even further.

Second Season Data Sub-set: Figure 4-12 and Tables 4-14 and 4-15 present the results of the SEM for the data sub-set of farmers interviewed in detail about their second season cropping practices together with a large number of variables independent of the particular season (e.g. age, institutional setting, knowledge on conservation issues).

Household wealth does not seem to play an important role in influencing the perception of the farmer on the satisfaction level of basic family needs or his choice of farm practices. Both linkages are not significant. Surprising is the negative sign for the linkage between wealth and basic needs satisfaction levels. It was hypothesized that the lower wealth level of the household the more the farm family will be worrying about how to obtain enough food and cash income to guarantee their survival.

Human capital in return impacts much stronger on these two endogenous variables; both linkages are significant. Especially its influence on how farmers perceive the basic needs satisfaction is quite strong. This leads to the conclusion that the more analytical skills the farmer possesses the better he is able to evaluate his overall family situation. Also for deciding what kind of technologies to use for reaching his overall production goals the amount of schooling and the farmer's age play an important role. In addition, also the physical conditions of the farm strongly influence the farmer's use of techniques, as the significant link between both variables shows.

On the other hand neither the knowledge of the farmer about erosion and conservation issues nor his institutional environment seem to impact on his intention of technology use. Both links are not significant and the coefficients are relatively small. It is very likely that the weakness of these linkages has to do with the fact that many farmers in the valley do not know very much about the degree of the existing erosion and soil fertility problems. Furthermore, only very few are familiar with soil conservation measures or are even utilizing them. Also the institutional set-up in the valley with respect to conservation issues is quite weak. The public extension service does not exist anymore, there is hardly any agricultural research done in the valley and only about two NGOs address erosion and soil conservation issues in their work.

Figure 4-12: Estimation results of the Structural Equation Model with latent variables for the data-subset for farmers interviewed about the SECOND growing season

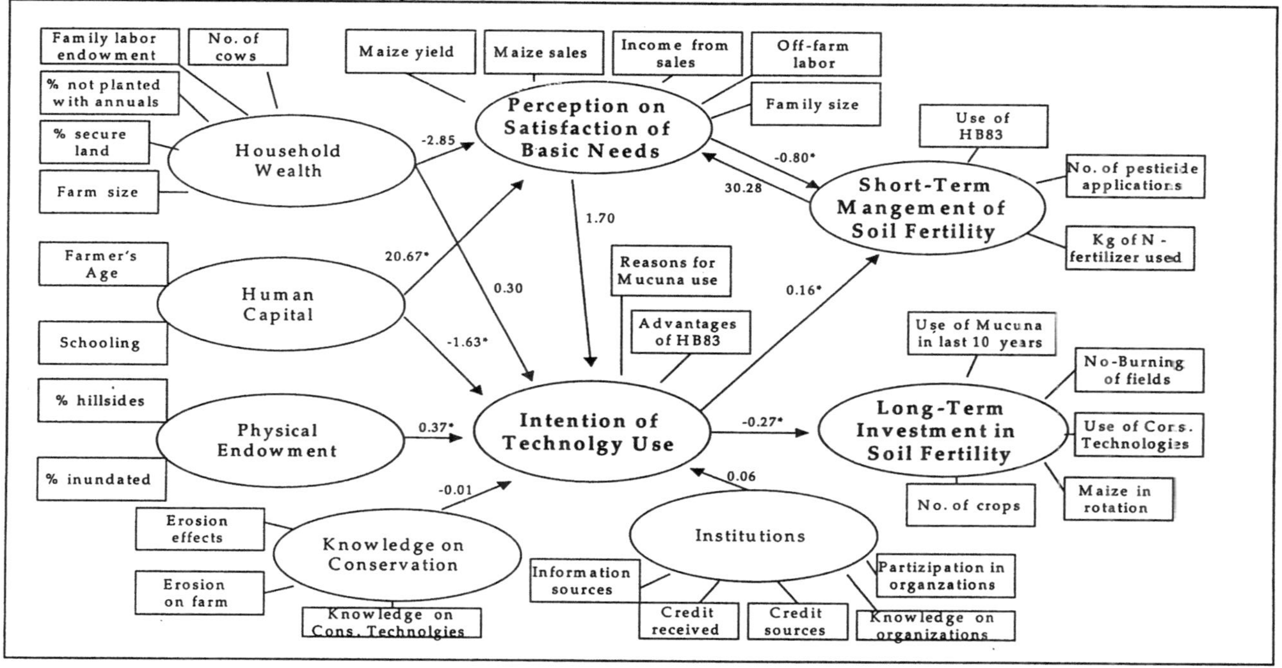

Source: Own Calculations

There also does not seem to exist a strong linkage between the degree to which the farmer is able to cover the basic requirements of his family and his technology choice.

Table 4-14: Results for the estimation of the path coefficients in the SEM calculated with the data-subset of farmers interviewed about the SECOND growing season

Latent Variables	Parameter estimate	Standard Error	P-level
Household Wealth → Perception on satisfaction of basic needs	-2.85	2.66	0.28
Household Wealth → Intention of technology use	0.30	0.21	0.16
Human Capital → Perception on satisfaction of basic needs	20.67*	3.58	0.00
Human Capital → Intention of technology use	-1.63*	0.23	0.00
Physical Endowment → Intention of technology use	0.37*	0.12	0.00
Knowledge on conservation → Intention of technology use	-0.01	0.07	0.85
Institutions → Intention of technology use	0.06	0.07	0.37
Perception on satisfaction of basic needs → Intention of technology use	1.70	0.00	-
Intention of technology use → Short-term management of soil fertility	0.16*	0.08	0.04
Intention of technology use → Long-term investment in soil fertility	-0.27*	0.11	0.01
Short-term management of soil fertility → Perception on satisfaction of basic needs	30.28	0.00	-
Perception on satisfaction of basic needs → Short-term management of soil fertility	-0.80*	0.18	0.00

n = 74, DF: 692, X^2: 1412.71, p-level: 0.00, Steiger-Lind RMSEA Index: 0.10, GFI: 0.54, AGFI: 0.49,
NFI: 0.28, Parsimonious fit index: 0.26
Source: Own calculations

Table 4-15: Results for the estimates of latent variables in the SEM calculated with the data-subset of farmers interviewed about the SECOND growing season

Latent Variable	Indicator Variable (unit)	Parameter Estimate	Standard Error	P- level
Household wealth	Farm size (ha)	0.55*	0.11	0.00
	% of secure land (%)	-0.18	0.12	0.12
	% without annuals (%)	1.00*	0.83	0.00
	Family labor endowment (adult eq.)	0.01	0.12	0.95
	Number of cows (headcount)	0.19	0.12	0.11
Human capital	Farmer's age (years)	0.18	0.13	0.16
	Schooling (years)	0.29*	0.13	0.03
Physical endowment	% hillsides (%)	-0.55*	0.11	0.00
	% inundated (%)	1.00*	0.08	0.00
Knowledge on conservation	Erosion effects 1 (1 to 4)	1.00*	0.08	0.00
	Erosion effects 2 (1 to 4)	-0.59*	0.11	0.00
	Erosion effects 3 (1 to 4)	-0.31*	0.11	0.01
	Erosion on farm (binary)	0.87*	0.09	0.00
	Knowledge on cons. technologies (binary)	-0.33*	0.11	0.00
Institutions	Information sources 1 (1 to 4)	0.78*	0.10	0.00
	Information sources 2 (1 to 4)	1.00*	0.08	0.00
	Information sources 3 (1 to 4)	0.17	0.12	0.14
	Credit sources (binary)	0.27*	0.12	0.02
	Credit received (binary)	0.24*	0.12	0.04
	Knowledge on organizations (binary)	-0.14	0.12	0.23
	Participation in organizations (binary)	0.02	0.12	0.90
Perception on satisfaction of basic needs	Maize yield (kg/ha)	-	-	-
	Maize sales (binary)	0.61*	0.18	0.00
	Income from sales (Quetzales)	0.38*	0.17	0.03
	Off-farm labor (binary)	-0.14	0.17	0.42
	Family size (headcount)	-0.21	0.17	0.23
Intention of technology use	Reasons for Mucuna use 1 (0 to 3)	-	-	-
	Reasons for Mucuna use 2 (0 to 3)	0.76*	0.07	0.00
	Reasons for Mucuna use 3 (0 to 3)	0.43*	0.10	0.00
	Advantages of HB 83 1 (0 to 2)	0.20	0.11	0.07
	Advantages of HB 83 2 (0 to 2)	0.44*	0.10	0.00
Short-term management of soil fertility	Use of HB 83 (binary)	-	-	-
	No. of pesticide applications (Number)	-0.51*	0.20	0.01
	Kg of N-fertilizer used (kg/ha)	-0.81*	0.21	0.00
Long-term investment in soil fertility	Use of Mucuna in last 10 years (binary)	2.40*	0.98	0.01
	No-burning of fields (binary)	-	-	-
	Use of cons. technologies (binary)	-0.60	0.46	0.19
	Maize in rotation (binary)	-3.30	0.42	0.47
	No. of crops (Number)	1.10	0.57	0.06

n = 74, DF: 692, X^2: 1412.71, p-level: 0.00, Steiger-Lind RMSEA Index: 0.10, GFI: 0.54, AGFI: 0.49,
NFI: 0.28, Parsimonious fit index: 0.26, Source: Own calculations

The calculated coefficient is not significant, though it is positive. It was expected though that this link would be stronger as it was hypothesized that the overall situation of the farm family would constitute a strong driving force in the choice between short- or long-term soil management measures. Especially since it seems that the inclination of farmers in the valley is more towards using soil management practices that increase productivity in the short run (CIs). The link between the variables 'Intention of technology use' and 'Short-term management of soil fertility' is significant and positive. This is a strong indicator of the farmer's mindset. In addition, for the link between the farmer's intention and the variable modeling the use of long-term soil fertility management measures a significant, negative coefficient was estimated. This demonstrates that farmers are not inclined to use this kind of practices when they have the option to raise soil fertility levels with measures that show quick results. This is another indicator for the rather myopic perspective of farmers under subsistence conditions.

An interesting relationship can be found between the satisfaction level of basic needs and the practices farmers use to increase short-term soil fertility. On the one side there exists a significant, negative relationship between satisfaction level and management practices. This means that if the farmer perceives that his family is lacking food and/or cash he will try to increase his productivity level through boosting soil fertility in the short-term. Nevertheless, the reversed linkage between the two variables is not significant, though it is positive as expected and the coefficient is relatively high. Therefore the use of short-term soil management practices does not automatically lead to increased food and cash availability.

Table 4-15 shows the results for the estimation of the relationships between the different indicator variables and their associated latent variables. The results will here not be discussed in detail, but it can be seen that for each latent variable significant indicators can be found.

According to the selected criteria the overall fit of the model needs improvement. The X^2 value for the model indicates a significant difference between the covariance matrices of observed data and the model. The Steiger-Lind RMSEA though is with 0.10 considered to be good. The GFI, AGFI, NFI all indicate a model fit that can be improved. This is the same for the Parsimonious fit index. To improve the overall fit in a next step of the analysis the non-significant indicator variables could be omitted and different ones chosen. Additionally non-significant paths could be omitted and the model in parts re-specified. Nevertheless, as a confirmatory approach had been chosen in this study no model modifications were carried out. The obtained results are therefore a first good indication into which direction further research can be carried.

First Season Data sub-set: As mentioned earlier, the SEM was also run with the data obtained by interviewing farmers about their first season cropping practices. The important season for taking the decision on planting Mucuan is the second season though because in the traditional Maize-Mucuna-System Mucuna is planted in this period. Nevertheless, today farmers also experiment with first season Mucuna planting. Additionally, for these farmers also data on their Mucuna planting in the second season were available.

Tables 4-16 and 4-17 and Figure 4-13 show the results for this model run. Here only the differences compared to the results of the second season data-subset will be discussed.

Comparing significant linkages between variables in the two data sub-sets, it can be seen that in the first growing season a different set of linkages becomes important. Only the variable 'Physical endowment' still significantly influences the 'intention of technology use'. Also the link between human capital and the technology choice of the farmer is still significant. New variables that impact in this season on the variable 'Intention of technology use' are household wealth and the perception on the satisfaction level of basic needs. This can result from the fact that the first growing season is mainly used to produce food for family consumption and refill food supplies after the dry season. Therefore now the ability of the farmer to buy food if stocks have become too low becomes important, which depends on the overall wealth situation of the farm family. Along these lines it is also interesting that now the link between short-term soil fertility management and the perception on the satisfaction of basic needs is significant. Surprising is nevertheless the negative sign of this linkage. It was expected that the better the farmer is able to boost short-term soil fertility, the better he is able to fulfill the basic food requirements of his family. As many of the short-term measures though require a cash investment in for example fertilizer and improved seeds material, farmers might feel that their overall level of basic needs satisfaction decreases in this period. Similar to the second growing season though neither the knowledge on conservation nor the institutional setting in which the farmer operates influence his technology choice. It is interesting to note that for this data sub-set the variable 'Intention of technology use' does not seem to influence significantly the use of either short- or long-term soil management practices. The positive sign for the link to short-term measures and the negative one for the linkage with long-term practices though point towards an overall inclination of farmers to put productivity before conservation goals.

Figure 4-13: Estimation results of the Structural Equation Model with latent variables for the data-subset for farmers interviewed about the FIRST growing season

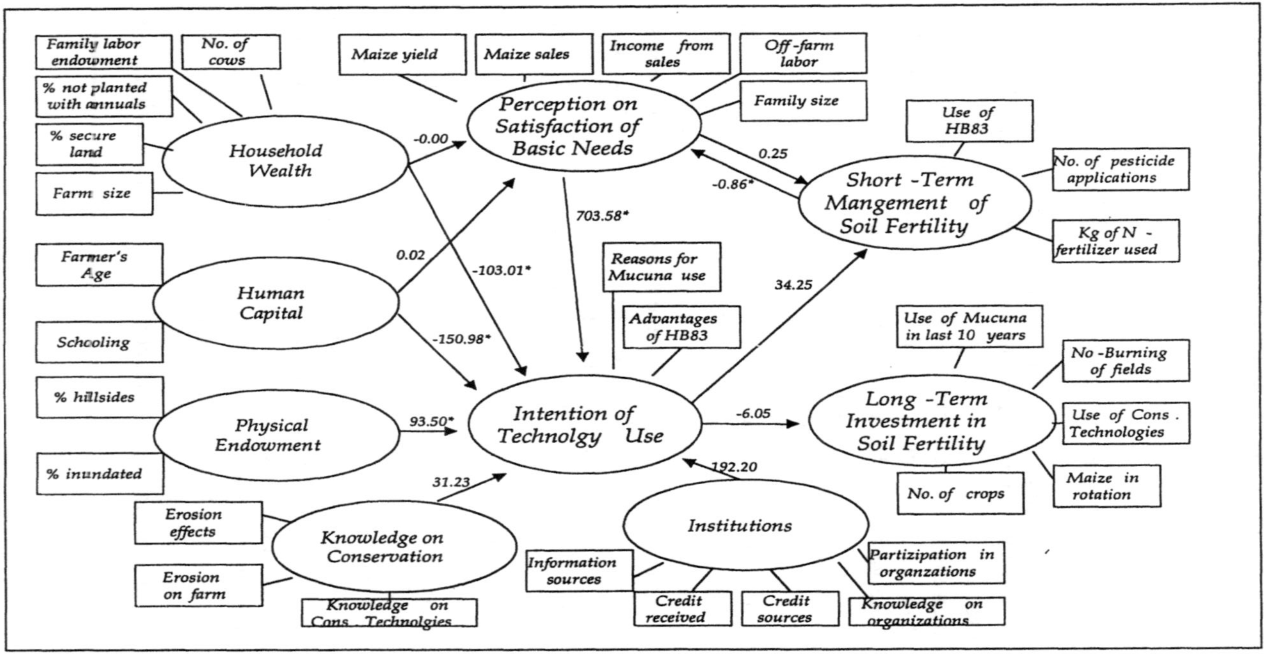

Source: Own Investigation

Table 4-16: Results for the estimation of the path coefficients in the SEM calculated with the data-subset of farmers interviewed about the FIRST growing season

Latent Variables	Parameter estimate	Standard Error	P-level
Household Wealth → Perception on satisfaction of basic needs	-0.00	0.13	0.99
Household Wealth → Intention of technology use	-103.01*	46.67	0.03
Human Capital → Perception on satisfaction of basic needs	0.02	0.17	0.89
Human Capital → Intention of technology use	-150.98*	59.47	0.01
Physical Endowment → Intention of technology use	93.50*	41.10	0.02
Knowledge on conservation → Intention of technology use	31.23	32.28	0.33
Institutions → Intention of technology use	192.20	0.00	-
Perception on satisfaction of basic needs → Intention of technology use	703.58*	249.65	0.01
Intention of technology use → Short-term management of soil fertility	34.24	138.37	0.81
Intention of technology use → Long-term investment in soil fertility	-6.05	24.65	0.81
Short-term management of soil fertility → Perception on satisfaction of basic needs	-0.86*	0.39	0.03
Perception on satisfaction of basic needs → Short-term management of soil fertility	0.25	0.14	0.09

n = 63, DF: 691, X^2: 1345.24, p-level: 0.00, Steiger-Lind RMSEA Index: 0.08, GFI: 0.56, AGFI: 0.50,
NFI: 0.36, Parsimonious fit index: 0.34
Source: Own calculations

Table 4-17: Results for the estimates of latent variables in the SEM calculated with the data-subset of farmers interviewed about the FIRST growing season

Latent Variable	Indicator Variable (unit)	Parameter Estimate	Standard Error	P- level
Household wealth	Farm size (ha)	0.89*	0.21	0.00
	% of secure land (%)	0.05	0.14	0.70
	% without annuals (%)	0.43*	0.15	0.01
	Family labor endowment (adult eq.)	-0.16	0.14	0.26
	Number of cows (headcount)	0.34*	0.15	0.02
Human capital	Farmer's age (years)	-0.57*	0.24	0.02
	Schooling (years)	-0.06	0.18	0.74
Physical endowment	% hillsides (%)	0.70*	0.18	0.00
	% inundated (%)	-0.83	0.20	0.00
Knowledge on conservation	Erosion effects 1 (1 to 4)	1.00*	0.09	0.00
	Erosion effects 2 (1 to 4)	-0.37*	0.12	0.03
	Erosion effects 3 (1 to 4)	-0.17	0.13	0.17
	Erosion on farm (binary)	1.00*	0.09	0.00
	Knowledge on cons. Technologies (binary)	-0.21	0.12	0.10
Institutions	Information sources 1 (1 to 4)	-1.00*	0.09	0.00
	Information sources 2 (1 to 4)	0.83*	0.10	0.00
	Information sources 3(1 to 4)	0.19	0.13	0.12
	Credit sources (binary)	0.22	0.13	0.08
	Credit received (binary)	0.20	0.13	0.10
	Knowledge on organizations (binary)	-0.08	0.13	0.53
	Participation in organizations (binary)	0.09	0.13	0.50
Perception on satisfaction of basic needs	Maize yield (kg/ha)	-	-	-
	Maize sales (binary)	1.64*	0.58	0.01
	Income from sales (Quetzales)	1.29*	0.50	0.01
	Off-farm labor (binary)	-0.39	0.36	0.29
	Family size (headcount)	0.11	0.35	0.76
Intention of technology use	Reasons for Mucuna use 1 (0 to 3)	-	-	-
	Reasons for Mucuna use 2 (0 to 3)	-0.21	4.13	0.96
	Reasons for Mucuna use 3 (0 to 3)	-0.36	4.30	0.93
	Advantages of HB 83 1 (0 to 2)	31.31	126,52	0.81
	Advantages of HB 83 2 (0 to 2)	23.69	95.76	0.81
Short-term management of soil fertility	Use of HB 83 (binary)	-	-	-
	No. of pesticide applications (Number)	-0.16	0.13	0.21
	Kg of N-fertilizer used (kg/ha)	-0.28*	0.12	0.02
Long-term investment in soil fertility	Use of Mucuna in last 10 years (binary)	0.73	0.40	0.07
	No-burning of fields (binary)	-	-	-
	Use of cons. technologies (binary)	-0.40	0.34	0.25
	Maize in rotation (binary)	-1.16*	0.57	0.04
	No. of crops (Number)	0.37	0.34	0.28

n = 63, DF: 691, X^2: 1345.24, p-level: 0.00, Steiger-Lind RMSEA Index: 0.08, GFI: 0.56, AGFI: 0.50,
NFI: 0.36, Parsimonious fit index: 0.34, Source: Own calculations

The overall model fit of the model with the first season data set is similar to the one with the second season data and therefore also needs improvement.

Summary of SEM results: There are a number of lessons that can be learnt from the SEM analysis. First, the physical conditions of the farm are of course of great importance for the farmers' technology choice. This is to be expected. Therefore any new practice will have to be adapted to the particular soils, topography and climatic characteristics of the area.

Second, most farmers are more inclined to use practices that boost soil fertility in the short-run, while the link between their technology choice and long-term practices is negative. This is to be expected under subsistence conditions. In addition, this inclination is influenced by the degree to which the basic needs of the farm household are covered.

The linkage between the institutional setting of the farmer and his technology choice are very weak. Neither his information sources, the membership in certain organizations nor his knowledge on credit sources that could be employed for obtaining the necessary means to introduce new practices or bridge production gaps play an important role. Also the general knowledge level of farmers on erosion and conservation issues is relatively low and does not lead to an inclination of the farmer to use more techniques oriented at conserving soil fertility in the long-run. These weak linkages can have severe consequences for the introduction of new practices in the future, especially if soil erosion becomes a more pressing problem in the valley as it already is.

Results describing the importance of certain structural factors, such as the farmer's level of wealth or human capital are mixed. Human capital, representing the experience a farmer has due to age etc., seems to be important. Wealth levels on the other side do not seem to be linked directly to the farmer's choice of practices.

4.6 Lessons learned

For quiet a long time, the Maize-Mucuna-System showed beneficial results for farmers in the Polochic Valley. It was and is well adapted to the prevailing maize production system and helps to boost yields. Therefore, it was and is used by a large number of farmers in the area, for whom this is the most important aspect of planting the 'Fertilizer bean'. In terms of labor, capital, and land endowments Mucuna fits the circumstances of small-scale farmers, as it does not require any substantial additional amounts of labor or capital for its introduction and maintenance. Also up to now enough land was available in the valley, so that farmers could rotate among their fields and leave parts in Mucuna fallow. Many farmers also continue using the system together with improved maize varieties, which shows that there are other factors than the introduction of new seed material involved in fostering changes in the overall cropping system or even leading to the abandonment of Mucuna planting.

One important reason for this development is the availability of arable land, which is slowly starting to decline in the valley. Quite a number of farmers today need to plant twice a year on their fields to produce sufficient output for satisfying the food and cash requirements of their families. Also the percentage of land that cannot be used in certain parts of the year due to inundation or unclear access rights is becoming important. This shows the shifts in resource endowments, particularly land, taking place in the area. Therefore fewer farmers are able to use the Maize-Mucuna-System with its one-term fallow period, despite farmers valuing the productivity enhancing aspects of the legume.

As explained in Chapter 3, understanding farmers' time horizons is important for analyzing the short- versus the long-term profitability of an environmental innovation. The stepwise abandonment of the Maize-Mucuna system demonstrates the rather myopic perspective of the small-scale farmers in the valley. If evaluating the system from a short-term perspective, Mucuna can be seen as preventing the intensification of land use. Land is perceived as becoming scarcer, which will guide technical change according to the "Induced Innovation Theory" towards land use systems that save land and the abandonment of the Mucuna system. Analyzed from a long-term perspective though Mucuna can be interpreted as a way of intensifying the use of land. Due to the fact that the long-term utilization of Mucuna helps to stabilize soil fertility, it allows the reduction of long-term fallow periods necessary in slash-and-burn systems to restore soil productive capacity. Hence, it is likely that this aspect of the Maize-Mucuna system will guide the direction of technical change towards a continuation of Mucuna use. Nevertheless, under the current institutional and economic settings in the valley and with the weak linkages between farmers' technology choices, institutions, and farmers' knowledge on erosion, reaching a transition in perspective from the short to the long-term will be difficult. Such a change though will be urgently required to halt the degradation problem in the valley.

The resource conservation aspect of the Maize-Mucuna-System does not seem to influence farmers' decision to use or continue using the legume. The investigated linkages between soil erosion and conservation issues and farmers' motivation to use short- versus long- term soil fertility management practices are quite weak, which is influenced by the insufficient knowledge that farmers have on soil erosion and its effects. And farmers show an overall inclination to manage soil fertility more according to short-term goals, leading to an increased use of commercial innovations (i.e. fertilizer, improved varieties). In addition, the institutional setting in the valley that could help farmers to introduce changes in their management practices through the provisioning of knowledge or credit is not very well developed. Therefore this institutional framework hardly influences the farmers' choice of technologies. Here the question arises how technical change, which is more oriented towards conservation of soil resources, can occur in the area. With time probably the demand for practices that also conserve the soil will become stronger, as soil fertility declines. According to the "Induced Innovation Theory" a continuous exchange between farmers and the research system that can provide new, location specific solutions is needed. But in the Polochic Valley the institutions to foster this change by researching and promoting new PERC technologies adapted to the changes in land availability are very weak and almost non-existent.

Another important issue that was pointed out already before should be mentioned again: Farmers in the Polochic Valley value the productivity enhancing aspect of Mucuna more than the conservation aspect. Different from the scientist's view of Mucuna, for whom the legume is mainly a conservation practice, farmers in the valley see it as a fertilizer replacement. Hence, the incentive for farmers to implement the technique lies intrinsically in the practice itself through the yield increase. From this the conclusion can be drawn that any conservation practices that is offered to farmers is more likely to succeed if it contains an important intrinsic incentive for farmers to adopt. Under small-scale farmers' circumstances this is very likely to be the productivity-enhancing component of the practice. If this incentive is coupled with an overall fit between farmers circumstances and the characteristics of the technology the likelihood of adoption is likely to be high. Therefore technology developers need to take these aspects into account when developing new practices aimed at adoption by subsistence farmers in developing countries.

5 Offering Soil Conservation Technologies to Farmers and Farmers′ Response: The Origin and Use of Soil Conservation Techniques in the County of Nueva Concepción, El Salvador

5.1 Introduction

The continuous development of new technologies and their adaptation to the specific conditions that shape farmers' livelihoods has been identified in the "Induced Innovation Theory" as one of the key factors of technical change. Technology developers and promoters have to be aware of farmers' resource constraints and inner and outer farm circumstances, which can present adoption obstacles. The process of technology development and adaptation needs to be flexible and responsive to changes in conditions of technology users.

In the case of Environmental Innovations (EIs), such as soil conserving technologies, factors hindering their use under small-scale farmers' conditions are manifold, as explained in the preceding chapters. Three characteristics of soil conservation technologies are considered particularly important for the case study presented in the following, as they might heavily influence technology development and farmers′ adoption behavior:

- The selection and implementation of soil conservation technologies require a lot of time and knowledge about the new techniques as well as insight into physical and biological interactions within the existing farming system.
- There often exists a substantial time gap between the implementation of soil conservation practices and the moment in which the farmer starts to receive benefits. Therefore, it is often difficult for farmers to see the direct benefits attached to adopting these practices.
- Many soil conservation technologies have public good characteristics, making their development less attractive for private sector commercial organizations.

Depending on the kind of technology to be promoted and developed further within an area a certain degree of adaptive research might be necessary to compensate for the complexity of the technology. Technology suppliers might also have to look at the agronomic and economic characteristics of the practices they choose to offer. In this context a number of questions arise:

- Who are the suppliers of soil conservation practices in Central America?
- How do they select the area to work in and how do they elicit the most pressing problems of their target population?

- Where and how do the organizations obtain information on soil conservation techniques and how do they process this information (adaptive research)?
- Who are the main actors in the technology development and adaptation process?
- How does the information flow/interactions between the organizations and farmers work?
- Do the technology developers consider technology characteristics when selecting the soil conservation techniques they plan to work with in a certain area?
- How is farmers' response, i.e. adoption, to the offered technology choices?

To answer some of these questions, a case study was carried out in the county of Nueva Concepción, El Salvador. The research area was chosen because of its major soil degradation problems combined with the predominance of small-scale agricultural systems. In addition, a substantial number of organizations were identified as working in the promotion of various soil conserving techniques. This case study is part of the efforts of the International Maize and Wheat Improvement Center (CIMMYT) and the Regional Maize Program (*Programa Regional de Maíz*), a network of Central American scientists working in maize research, to better understand the factors influencing the development and adoption of soil conserving technologies in the region.

Objectives of the study are the following:

- To investigate who works in the promotion of soil conservation techniques in the county and if there are differences between the different technology suppliers.
- To study selection criteria for soil conservation techniques and the process of transferring them to farmers.
- To measure adoption levels of the offered soil conservation practices and investigate farmers sources of information for these techniques.

In the study technology suppliers and selected farmers were interviewed to test the following general hypothesis and two sub-hypotheses:

General Hypothesis:

- The characteristics of the soil conservation techniques offered to farmers only partly match those demanded by farmers.

Sub-Hypotheses:

- Short- or medium-term profitability of the supplied technologies has not been given sufficient consideration when selecting the promoted techniques.
- There has not been sufficient adaptive research in the area to tailor the selected techniques to particular farmers' circumstances.

The case study presented aims at piecing together a first small picture of the system that led to the promotion and diffusion of a few selected soil conservation technologies in the county. To the knowledge of the author this kind of investigation has not been carried out yet in other parts of Central America. Special attention is given to zero tillage systems with no-burn practices and crop residue management. This particular technique is considered a productivity enhancing and resource conserving (PERC) technology because soil erosion is effectively reduced through this practice by providing a protective cover for the soil during rainfall. At the same time yields can be stabilized or even enhanced by providing organic matter to the soil, thus reducing the water loss due to evapotranspiration.

5.2 Methodological Issues

Representatives of 14 organizations working with soil conservation practices in Nueva Concepción were interviewed, according to what farmers had given in a pre-survey as their source of information for soil conservation practices. In addition, some institutions had been mentioned by some of these organizations as collaborators for their work with conservation techniques. Farmers mentioned two additional institutions, whose representatives could not be interviewed because they were either not working any more in the area or it was not possible to meet the representative during the time of the survey. The interviews touched on the following issues (survey in Annex 2):

- Criteria for the selection of the promoted technique(s)
- Level and intensity of adaptive research before and during the promotion of the selected technique(s)
- Participantion in the adaptive research process
- Methods used to reach the adoption of the promoted technique(s)
- Institutional origin of the information about the promoted technique(s)
- Information flow within the organization

In addition, a survey of 76 randomly chosen farmers was carried out in the county. The farmers were chosen from each of the ten districts that comprise the county. No farmers living in the county capital were interviewed. Thus in total 0.4% of the inhabitants of the county were surveyed. The number of interviewed farmers per district was determined according to the number of inhabitants per district. The following issues were investigated (survey in Annex 3):

- Adoption rate of different soil conservation technique(s)
- Reasons for adopting these technique(s)
- Maintenance of the technique(s)
- Reasons for non-adoption or abandoning the promoted technique(s)
- Information sources about conservation technique(s)

- Incentive structure and/or subsidies given to farmers to facilitate the adoption of conservation techniques

As the number of interviewed organizations and farmers is relatively small, the data were analyzed using descriptive statistical methods only.

5.3 Description of the study area

In the following a brief description of the study area and the current agricultural production system is given.

5.3.1 General description

The county of Nueva Concepción lies at the foot of a mountainous region in the northern department of Chalatenango, which borders Honduras (Figure5-1). It comprises a total area of 258 km^2 and is divided into ten districts.

The climate of the region can be described as dry subtropical with an annual precipitation of about 1500 mm/year and a medium annual temperature of around 22° C. The rainy season lasts from May until October/November, though a small period (canícula) of about 20 days with relatively little rain can be observed in July/August (CENTA/MAG and IICA-Holanda/LADERAS 1996).

The area lies in the transition zone between the humid subtropical forest, found in the east of the county, and the dry subtropical forest in the middle and the west. Altitudes range from 325 to 750 masl. The county comprises three main agro-ecological areas: In the south and southeast the alluvial plains of the Lempa River with medium to high soil fertility constitute about one fourth of the area. Towards the west and north of the plains hillsides with slopes between 15-50% give way to a mountain chain with inclinations of over 50%. Thus about 70% of the total area of the county consists of hillsides with restrictions for agricultural production. Additionally, the soils in this part are loamy to sandy, relatively shallow, of moderate to low soil fertility and with a very high stone content. Their susceptibility to erosion is described as high (Mercado and Rodriguez Sandoval 1996). In a Master Plan for the development of the department of Chalatenango designed by the FAO and the project "PROCHALATE" 45% of the county of Nueva Concepción was classified as land with potential for crop production, 38% as land that can only be used for pastures or perennial crops and forest, while17% should not be used in any kind (IICA-Holanda/Laderas 1995).

Figure 5-1: Map of El Salvador and the study region

Source: University of Texas, Map Collection

The population of the county (about 28.000 inhabitants; Censo Nacional de Población 1992, in: IICA-Holanda/Laderas 1995) is of indigenous origin (Chortis Mayans), but today it is mixed with Ladinos and Europeans. Population density varies throughout the county: Two of the ten districts have more than 100 inhabitants/km^2, while six other districts count with more than 60 inhabitants/km^2. The highest density with more than 300 inhabitants/km^2 can be observed in the little town of Nueva Concepción, which lies in the middle of the county (Mercado and Rodriguez Sandoval 1996).

Agriculture is the most important activity for the population. In a survey carried out in 1994 by the project "PROCHALATE" it was found that 71% of the surveyed people lived in the rural zone and 88% were involved in agricultural activities (IICA-Holanda/Laderas 1995). Maize, rice, beans, sorghum, sugar cane and cattle are the most important agricultural products of the county. Nevertheless, the production does not seem to be sufficient for many farmers as during the dry season from December to March over 60% of the rural population, mainly men, migrate to other departments to work in the coffee or sugarcane harvest or in infrastructure projects (CENTA/MAG and IICA-Holanda/Laderas, 1996).

The majority of farmers do not own the land they cultivate. The "PROCHALATE" survey states that 45% of the interviewed farmers have to rent their land; almost all land renters (41% of all farmers) can afford to rent only less than 2 ha. 21% of the farmers belong to agricultural co-operatives, which were created due to an agrarian land reform as part of the peace agreement in 1992. The 8% of farmers that own more than 7 ha possess about 62% of the agricultural area. These general conditions were confirmed by a survey of the public research organization "CENTA" carried out in 1996, in which 53% of the interviewed farmers were identified as land renters and 39% as landowners. Thus land scarcity is an obvious problem for many farmers, which also has consequences for the introduction of soil conservation techniques (IICA-Holanda/Laderas 1995).

Soil erosion is a very important and visible problem throughout all hillside areas of the county. In the above-mentioned survey of the project "PROCHALATE", for example, 85% of the interviewed people state that they know that the deforestation rate in the county is too high. 75% mention that they face erosion problems on their farm, but only 21% use any kind of soil conservation practices. In 1996 about 70% of the farmers in hillside areas stated in the above-mentioned CENTA survey that they use the traditional practice of burning crop residues before the first rains, leaving the soil very susceptible to erosion (IICA-Holanda/Laderas 1995).

5.3.2 The farming system

As everywhere in Central America also the farmers of Nueva Concepción distinguish between two growing seasons: The first one (*primera*) starts with planting during the first rains in May and ends with the short period of drier days in July/August (*canícula*). The *segunda* then begins once the rains start to pick up in frequency again in September (months vary a bit according to location). Accordingly farmers grow up to two crops per year, which in traditional systems are usually two different crops. In more commercialized systems farmers tend to use more mono-cropping practices.

The alluvial plains in the county, created by the Lempa River, have a high potential for agricultural production. Thus they are mainly used for mono-crop production of cash crops like rice, sugar cane and vegetables as well as for pastures for raising cattle. Additionally some basic grains, like maize and sorghum, are cultivated in mono-cropping systems.

In the hillsides maize, beans and sorghum are the traditional crops, cultivated mainly by subsistence farmers with less than 5 ha. Various systems of intercropping or rotating the different crops exist. The most widely system used is the planting of maize, the main subsistence crop in El Salvador and by far the most important crop grown in the county, in May/June followed by sorghum between 30 to 60 days after maize planting in-between the maize rows. Both are sometimes also grown alone in the first season and then the field is left in fallow until the following year. Due to land scarcity this system is relatively rare nowadays. Beans are either intercropped with maize of the first growing season or sometimes grown in monoculture in the second season. A special crop of the hillside areas is sesame that is intercropped with maize. Pastures for extensive cattle production and some patches of secondary forest can as well be found.

The mountains are either used as natural pastures for cattle or they are covered with forest. Nevertheless, some maize/beans/sorghum fields can sometimes be found even on very steep slopes.

5.4 Description of organizations promoting soil conservation practices in Nueva Concepción

In this section an overview will be given about the organizations that work in the promotion of soil conservation practices in the county of Nueva Concepción to get a clearer picture of how conservation practices are supplied to farmers and whether different types of institutions employ different ways of reaching their clients. In the following section farmers' response to the promotion strategies of these institutions will be investigated.

Representatives of 14 organizations were interviewed, which is considered to be a high number for this small county. This might have something to do with the severity of the erosion problem in the area. Another reason could be that the county was severely affected by the civil war, so that there are now many organizations present in the area working in the mitigation of the war damages. The institutions then sometimes also work in promoting the use of sustainable agricultural practices.

In this section, first an overview of general characteristic of the investigated organizations will be given. For this purpose the organizations are grouped into three categories. Then the process of selecting the soil conservation practices employed by each organizational type is examined in further detail. In a last step the techniques used to promote the selected practices will be described.

5.4.1 General description of the investigated organizations

In Nueva Concepción there are three different types of organizations working in the promotion of soil conservation practices:

- NARS (National Agricultural Research System), which are the publicly funded El Salvadoran research and extension agencies. In the county these are the National Research and Extension Organization CENTA, the National Department for Renewable Natural Resources (Dirección General de los Recursos Naturales Renovables) and the National Department for Community Development DIDECO (Dirección de Desarrollo Comunal).
- Non-Government Organizations (NGOs), which are usually privately funded and are not related to any government or local official organizations. They operate either on a national or an international level or both. In Nueva Concepción the following NGOs were present: CONAMUS, Visión Mundial, FUNDANUEVA, CORDES and TECHNOSERVE.
- Other (foreign) Organizations, which are present in the area in form of specific projects or programs. Projects are designed to alleviate specific problems of the area in a certain period of time, after which the project should have made itself obsolete. In the county there are national as well as foreign projects, which either work specifically with soil conservation issues or are more geared towards poverty alleviation in general. These are IICA-Holanda/LADERAS, PROCHALATE and GECA. Programs are designed on a long-term basis and work on a national or regional level with the purpose of addressing a certain problem or field of interest that might be common to

various areas (e.g. soil erosion in Central America, maize research in Central America). In the county there are two programs present: One is national, executed by an international consultancy firm (PAES/Abt. Windrock). The second is a program for sustainable agriculture financed by the Swiss Government, which works in three Central American countries (PASOLAC). Additionally, there is one publicly funded foreign development agency present in Nueva Concepción, in this case the American Peace Corps.

Figure 5-2: Types of Organizations working in Soil Conservation in Nueva Concepción

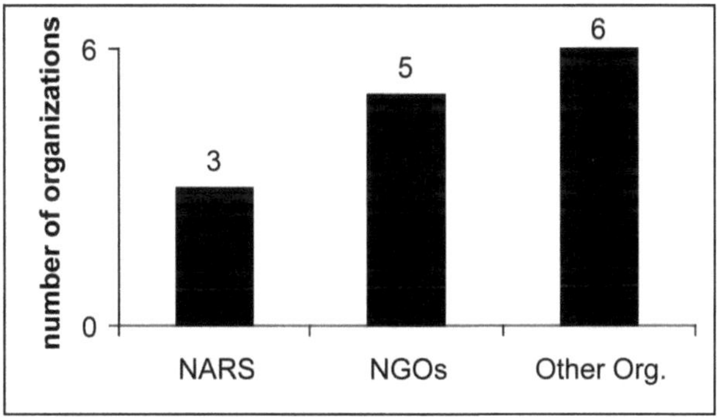

Source: own investigation, N = 14

As can be seen from Figure 5-2 most of the organizations working in the county belong to the type of "Other Organizations". The investigated institutions had different reasons for choosing to work in Nueva Concepción. In Table 5-1 can be seen that only some of the institutions, like the "Other Organizations", mention environmental issues such as the high proportion of hillsides and their associated problems or the protection of water sources as a reason for selecting the area. The agricultural production potential and poverty alleviation by improving the production opportunities of small-scale farmers are also important issues for these organizations as their mission often is geared more towards specific problems in agriculture and natural resource management. Some of these institutions even have quite elaborate criteria for their selection process that include environmental and technology indices and social indicators (e.g. IICA- Holanda/ LADERAS). For the NARS, the collaboration with other organizations present in the area and the existing infrastructure is an important issue, while the NGOs mainly came to the county for reasons not directly related with the agricultural or environmental problems. This fact will become clearer if we look at the range of general activities of the investigated institutions. The NGO work with soil conservation practices is more often part of other activities like credit or land transfer programs. Or they are there to deal with problems connected with the El Salvadorian civil war, which affected the area quite severely. It is interesting to note that some of the NGOs also came to the area or to specific communities in the county on request of these villages.

Table 5-1: Reasons given by the investigated organizations for working in Nueva Concepción

Reason for working in Nueva Concepción	**NARS (3 org.)**	**NGOs (5 org.)**	**Other Org. (6)**
High proportion of hillsides			++
Protection of water sources		+	
Production potential of the area	+*		+
Improvement of production possibilities of small-scale farmers			++
Collaboration with other organizations already present in the area	+		++
Accessibility and existing infrastructure	++		
On request of different communities		++	
Part of peace agreements after civil war		+	+
Credit program		+	
Part of land transfer program		+	
Participation in a competition		+	+

* + = signifies 1 institution of this type (i.e. if there are 3 NARS up to 3 + in this category can be given)
Source: own investigation, N = 14

The different types of organizations have different ranges of operation. While the public research and extension organizations cover the whole county, the other institutions work on a smaller scale, more in specific districts. This is especially the case for NGOs, which usually concentrate on about 2 districts. Figure 5-3 is a bit misleading though as it only shows averages. There are some projects as well as programs that work in the whole county while others work in a more specific area.

Figure 5-3: Number of districts in Nueva Concepcion, in which the different types of organizations worked in 1999

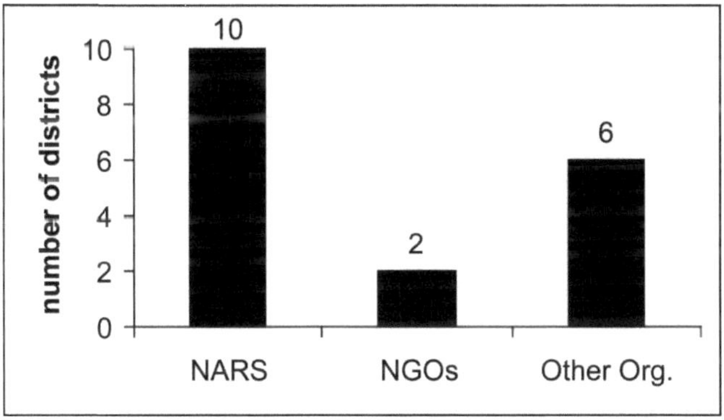

Source: own investigation

The investigated institutions also started their work in the county at different times. While all NARS are present in the area for more than 10 years this is quite different for the other institutions (Figure 5-4). Most NGOs and "Other Organizations" came to the region after the end of the civil war and have worked in the county no longer than six years.

Figure 5-4: Time period that investigated organizations work in Nueva Concepcion

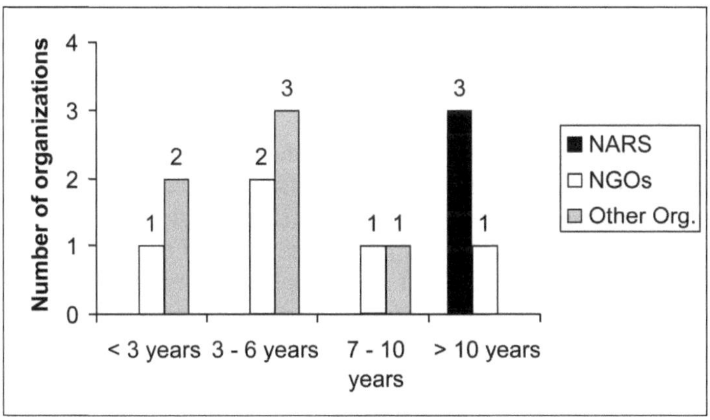

Source: own investigation, N = 14

The different scale of activities can also be seen in the number of people that work in the different institutions (Figure 5-5). The NARS have the highest number of people working in El Salvador, followed by NGOs. "Other Organizations" in general have a relatively low number of people working directly for them because they often collaborate closely with various institutions in their sites of operation. These collaborators usually pursue the actual work with the farmers while the people of "Other Organizations" coordinate the efforts in different areas and organize information exchanges and training workshops.

Figure 5-5: umber of people working in the different types of institutions in El Salvador, in Nueva Concepción and within Nueva Concepción in the promotion of soil conservation practices

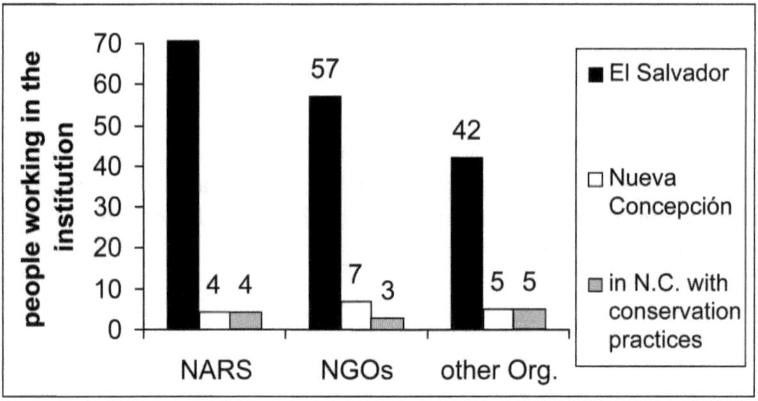

Source: own investigation N = 14

All institutions do not work exclusively in the promotion of soil conservation practices. This can be seen especially with NGOs, where the number of people working in the area differs quite a bit from the number of people working with soil conservation issues. Additionally let us look at the range of activities of the organizations (Table 5-2). NGOs in general are not only working in the field of Natural Resource Management (NRM), but they cover in addition a relatively wide range of other activities, like health issues, legal advice etc. They also focus particularly on building human capital and organization building skills within communities, which are all important issues for strengthening the self-help capacities of the rural communities. NRM and agriculture seems to be a part of this work, but not its main focus. This is quite different with NARS, which have a relatively straightforward agenda in research and technical assistance in agriculture and related issues. This of course also includes courses and organization building capacities, but these activities seem to be only a small part of the work. Also "Other Organizations " focus on education, training and technical assistance for farmers and human capacity building and strengthening of organizational skills is an important aspect of their work. Interesting to note here is also that one of the Programs working in Nueva Concepción is the only institution that mentioned the evaluation of technologies as part of their work agenda. We will come back to this point later, when we look at the research carried out by the investigated organizations.

Table 5-2: Activities carried out by different organizations in Nueva Concepción in 1999

Type of activity	NARS (3 org.)	NGOs (5 org.)	Other Org. (6)
Conservation issues and natural resource management	+++*	+++++	++++++
Evaluation of different conservation technologies			+
Courses/Education in agriculture		+	+++
Technical assistance (different crops, cattle, IPM, nurseries, diversification etc.)	++	+	++++
Agricultural commercialization	+		
Strengthening of organization building capacities	+	++++	++++
Human capacity building/educational programs	+	++	++
Health		+	
Social issues (education, child care, infrastructure etc.)	+	++++	+
Religious education		+	
Credit programs		+++	+
Legal advice		+	
Micro-Enterprises			+

* + = signifies 1 institution of this type (i.e. if there are 3 NARS up to 3 + in this category can be given)
Source: own investigation, N = 14

But how do the investigated institutions decide on the topics they think are important for farmers of the area? How do they elicit the most pressing problems of their target group, the small and medium size farmers of the county? And which process leads to the selection of the solutions and practices that will be promoted by the organization? These are important questions with respect to the supply of EI technologies.

5.4.2 The selection process of soil conservation techniques to be promoted

The first step of a particular organization to begin the work in an area and to understand the most important problems of the target group is usually to elicit a list of problems specific to the area. As farmers' demand and the involvement of the target group has become an important issue in rural development work, all institutions used participatory methods, like Rapid Rural Appraisals, etc. (Table 5-3). For this, either workshops were organized within certain communities or members of the institutions visited specific farms. Other types of information acquisition like formal and informal surveys were additionally employed by the NARS. Conducting surveys is the traditional method used by these institutions to gather information on the situation of their target population. Interesting is that some of the NGOs employ this rather traditional method as well and additionally rely on collaborating institutions and the knowledge of their technical personnel. As mentioned earlier, "Other Organizations" mainly rely on interactive methods, but might also consult other sources, like the literature or other institutions.

The request for help in particular problems can also come from a community, as the case of two NGO shows. Here it is interesting to note that the community went to these organizations and not to a governmental one, which would have been the case traditionally. Nowadays development experts evaluate the chances for a real change within the community as being higher, if requests for help come from the people themselves. Here a modern concept of interaction between development agencies and their clientele is followed, in which the institution responds directly to farmers need for solutions to particular problems and keeps in close contact with the respective communities through continuous meetings. Nevertheless, this concept implies a relatively high level of organization within the community, which is one reason why many institutions work in building organizational capacity.

Table 5-3: Methods used by investigated organizations to elicit the main problems of farmers in Nueva Concepción

Methods to elicit main problems of farmers in the area	**NARS (3 org.)**	**NGOs (5 org.)**	**Other Org. (6)**
Participatory appraisal with the community	+++*	+++	++++
Participatory appraisal with selected individual farmers	++	+	+++
Participatory appraisal carried out by collaborating organizations		+	+
Workshops with key informants			+
Informal survey	++	+	
Formal questionnaire		+	
Literature review (incl. review of surveys carried out earlier)			++
Knowledge of technical personnel that comes from N. C.		+	
Community seeks help of organization		++	
Regular community meetings		+	+

* + = signifies 1 institution of this type (i.e. if there are 3 NARS up to 3 + in this category can be given)
Source: own investigation, N = 14

After eliciting the main problems of their target population, the investigated institutions had to work on solutions to these problems. In the case of soil erosion, a wide range of EIs was offered to farmers (Table 5-4). They comprise techniques belonging to all three categories of soil conservation practices (see Chapter 2): Physical structures, Changes in Soil Management and Complex Management Systems. In the category of Changes in Soil Management the two main PERC technologies, no-burning systems and cover crops, can be found. Furthermore, the investigated organizations promote some practices that are related to general resource management, like establishing nurseries or even doing some kind of land use planning for individual farms. Furthermore, the organizations work with some new agricultural practices such as crop diversification, pesticide management or the establishment of organic fertilizer pits.

If we look at the specific conservation techniques, we see that live barriers made from different crops are the most widely promoted practice, followed by drainage ditches and no-burning practices. Dead barriers and agro-forestry systems are also quite popular with the conservation organizations, while other physical structures or other improved soil management practices play a more marginal role.

All investigated institutions cover a similar range of conservation technologies and offer a similar range of practices but they put a different emphasis on specific technology types. Figure 5-6 shows the kind of technology classes the different institutional types mainly work with. NARS put the main focus on Physical Structures and Changes in Soil Management, while General Resource

Table 5-4: Conservation and agricultural practices promoted by the investigated organizations

Promoted Conservation and Agricultural Practices	NARS (3 org.)	NGOs (5 org.)	Other Org. (6)
A. Physical Structures			
• Individual terraces	+*	++	
• Bench terraces		+	+
• Dead barriers of stone or straw	++	+	++
• Drainage ditches	+++	++	++
B. Changes in soil management practices			
• No-Burning practices with crop residue management	+	++	+++
• Live barriers with grasses, *Gliricidia sepium*, *Cajanus cajan* or pineapple	++	++++	++++
• Contour tillage (curvas a nivel)	+	+	++
• Contour ridges (callejones)			+
• Legume intercropping/Cover crops		+	+++
• Living fences	++		+
C. Introduction of a complex system			
• Agro-forestry	++	++	+
• Tree planting/Reforestation	+	+	
D. Practices related to natural resource Conservation			
• Nurseries	+	+	++
• Land use layout design of the farm			+
• Protection/management of water wells		+	
E. Agricultural Practices			
• Planting of fruit trees		++	
• Improved pastures	+		+
• Crop diversification			++
• Pesticide management	+		
• Use of organic fertilizers		+	
• Horticultural crops	+	+	+
• Organic fertilizer pits		+	

* + = signifies 1 institution of this type (i.e. if there are 3 NARS up to 3 + in this category can be given)
Source: own investigation, N = 14

Management Practices are hardly promoted at all. With NGOs the focus shifts away from Physical Structures to Changes in Soil Management practices and new Agricultural Practices. General Resource Management also becomes more important as some of these institutions work in the protection of water wells. This trend becomes even stronger with the "Other Organizations", which put their main emphasis on Changed Soil Management and an overall better management of agricultural and natural resources.

Figure 5-6: Shares of activities for different types of soil conservation practices offered by the investigated institutions in Nueva Concepción

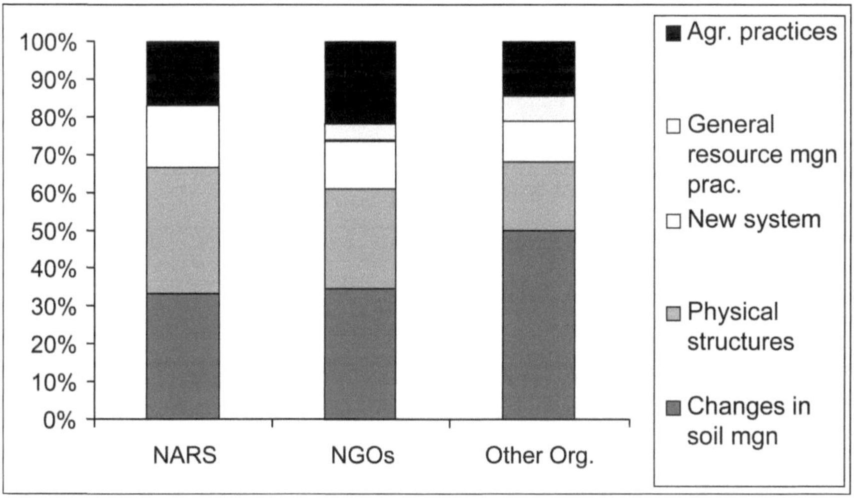

Source: own investigation, N = 14

It can be hypothesized that these differences in focus are a result of the selection process employed by the organizations to decide on the techniques they promote. In Table 5-5 the selection criteria of the different organizations can be seen in detail. NARS do not seem to have many specific selection criteria for choosing the technologies they promote. They try to be as responsive as possible to what was decided in community meetings with farmers and other organizations or to farmers' suggestions, but they do not seem to look at technology characteristics in particular. One NARS representative on the other hand mentioned that sometimes internal reasons, such as the help they can get from the ministry for working with certain techniques, might determine offered technology packages. This might be one reason for the still relatively strong promotion of physical structures mentioned above.

These technology types can show good results in reducing soil erosion but are often more difficult to adopt under small-scale farmers circumstances (see Chapter 2, e.g. large initial capital investment, profitable only after extended period of time). Farmers' adoption rates in the county for these techniques are examined in a later section.

Table 5-5: Criteria of investigated organizations in Nueva Concepción for their selection of soil conservation techniques

Criteria for the selection of soil conservation techniques	NARS (3 org.)	NGOs (5 org.)	Other Org. (6)
A. Economic criteria			
Technique does not require a large investment		+++	
Technique does not require high labor input		+	
Technique has a favorable cost-benefit profile, also in the short-run		+	++
Technique is easy to adapt to current production system			+
Technique is easy to understand/close to farmers' experiences		+++	+
Fast impact of technique		+	+
B. Ecological and agronomic criteria			
Climatic aspects		+	
Good protection of water sources		++	
Good for fighting soil depletion			+
Good soil cover		+	
C. Other criteria			
Selection according to work plan developed with farmers/county	++*	+	+
Suggestions of farmers	+	+	++
Suggestions of other organizations			+
Multiple use of technique possible			+
Easy for org. to work with technique (help of ministry etc.)	+		

* + = signifies 1 institution of this type (i.e. if there are 3 NARS up to 3 + in this category can be given)
Source: own investigation, N = 14

NGOs on the other hand have a wider range of specific selection criteria. Interesting here is that the economic characteristics of the promoted techniques were mentioned more often than their agronomic particularities. The promoted techniques should especially not imply a large investment for farmers and should be easy to understand. These criteria are consistent with circumstances of small-scale farmers in Central America, such as capital constraints and a relatively low education level. In addition, as some NGOs work in the protection of water sources, they are interested in specific techniques for this purpose and then ecological/agronomic reasons become important for their technology choice. Thus many NGOs promote, in addition to physical structures, more easily to implement changes in the farmers' soil management system and techniques that are more related to general natural resource management. Nevertheless, the promotion of physical structures with the above mentioned problems for small-scale farmer adoption does not completely correspond to the selection criteria given by institutional representatives. This might have something to do with the fact that only one of the NGOs mentioned that they consider farmers' suggestions and plans made with the communities as a criterion in their technology selection process. On the other hand this could also mean that some NGOs do not really look at the characteristics of the practices they offer. Thus they have a good understanding of farmers' circumstances and priorities when looking for new technologies, but they do not consider or know enough about the other side of the coin, the technology characteristics. To investigate this point a bit further on we will look at the sources of information about the promoted technologies that NGOs and the other institutions use.

For the "Other Organizations", economic criteria for selecting the techniques offered to farmers also play an important role. Here it seems that especially the profitability of the practices for farmers and a rapid impact on yields, but also their compatibility with the existing production system and farmers' experiences are considered. Of course the technique also needs to have a substantial impact on reducing erosion and related soil nutrient depletion. Similar to NARS the "Other Organizations" also try to incorporate farmers' and community suggestions and perspectives into their selection process. Thus they consider farmers demand for technologies, but they also seem to look at the traits of the possible options, which can be supplied to farmers.

The picture we find here is consistent with the answers given by the representatives of the investigated institutions on who is involved in the final decision about the range of conservation techniques offered (Table 5-6). As mentioned before, NARS want to incorporate farmers' and community demand for conservation techniques into the selection of offered practices. Thus their decision is taken either by them together with the farmers or by the community. Only one of the NARS still mainly relies on its technical personnel. In the "Other Organizations" the situation is similar. As the decision is mainly taken in close co-operation with the farmers, the technical staff of the institutions can play a decisive role in laying out the advantages and disadvantages of particular practices. With NGOs it is mainly the technical personnel that decide on the technology packages. Nevertheless, other opinions are seen as a decision-making aid, though the main decision power lies more at the organizational level than with farmers or with the communities when compared to NARS and "Other Organizations". Thus we can see a different emphasis on farmers' participation in this decision process and it would be expected that this would also lead to a different range of selection criteria.

Table 5-6: Final decision-makers for investigated organizations in Nueva Concepción about soil conservation techniques they offer to farmers

Final decision-makers for selection of promoted soil conservation techniques	NARS (3 org.)	NGOs (5 org.)	Other Org. (6)
• Technical personnel of organization	+*	++++	++
• Organization together with farmers	+	+	++++
• Organization together with other institutions		+	+
• Communities	+	++	

* + = signifies 1 institution of this type (i.e. if there are 3 NARS up to 3 + in this category can be given)
Source: own investigation, N = 14

In general, the selection criteria given by the different organizations are congruent with the soil conservation techniques promoted. Interesting is here to note that NGOs and "Other Organizations" have mentioned mainly economic criteria, which shows that they are aware of farmers adoption problems with conservation technologies and some of the important reasons for it. In addition, it looks like that NARS do not have specific selection criteria but rely on general knowledge about techniques and farmers' perception of them.

If we look at information sources used by the investigated institutions concerning the promoted conservation techniques, we see (Table 5-7) that the staff of the NARS working in Nueva Concepción mainly rely on course and training material put together by their headquarter or by other organizations. This is also the case with "Other Organizations". But in contrast to NARS these institutions together with NGOs additionally stress the professional knowledge of their staff. Furthermore, they try to strengthen the staff's own initiative to exchange information with other organizations, do literature reviews and learn from farmers' and community experiences. Thus in both of these organizational types staff seems to be more actively involved in the search for information. This might be one of the reasons why these organizations also have more specific technology selection criteria than NARS.

Research carried out by the investigated institutions does not really seem to be a source of information (Figure 5-7). One NARS, which is the official agricultural research agency of El Salvador CENTA, carries out formal investigations in most of the promoted techniques. A small part of this research is executed in Nueva Concepción, but the majority is done in other regions as part of the overall research activities of the organization. In total nine people work with this NARS in the county. For them the work with conservation practices is just one of their activities.

Table 5-7: Information sources of investigated organizations for promoted technologies

Information sources for promoted soil conservation techniques	NARS (3 org.)	NGOs (5 org.)	Other Org. (6)
• Material & courses made by other organizations	++*	++	
• Training courses & material prepared in HQ of organization	++	+	+++
• Information exchange with other organizations	+		++
• Academic training of staff			++
• Work experience of staff		+	+
• Excursions		+	
• Literature reviews		+	++
• Experiments with farmers			+
• Through the communities		+	

* + = signifies 1 institution of this type (i.e. if there are 3 NARS up to 3 + in this category can be given)
Source: own investigation, N = 14

One of the NGOs and three of the "Other Organizations" are involved in so-called "Farmer Experimentation". Here farmers do some kind of informal research on new technologies, often in close collaboration with the technical staff of the organizations. This research can help very much to adapt these new practices to specific local conditions, but unfortunately results are often not well documented and research procedures not standardized. This makes it difficult to disseminate experiences and compare them with each other and with results from other areas. Here already a few results of the survey carried out with farmers of the county on their use of soil conservation techniques should be mentioned. Farmers were asked if they changed anything or experimented with the practices they are using. Not one farmer did. Thus the number of farmers involved in "Farmers' experiments" is relatively small and farmers in general do not seem to like experimenting on their farm. In addition, less than one fourth of the interviewed farmers (24%) know of or participate in farmers groups working on spreading awareness of the land degradation problem and its possible solutions. These farmers' groups could be one possibility to stimulate the innovative potential of farmers in the county.

Figure 5-7: Number of investigated institutions doing research on the promoted soil conservation techniques

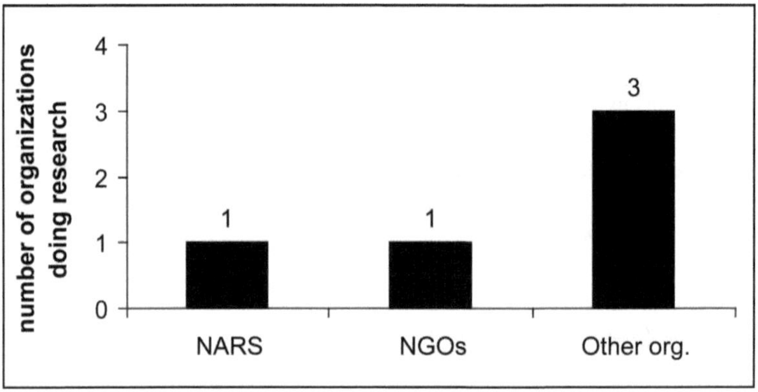

Source: own investigation, N = 14

If the number of researchers working in the region is compared with the number of people involved in the promotion of practices, we see that 43 % of the people are engaged with one of the organizations that also have some research activities. Nevertheless, this does not mean that all of them are also involved in research projects. The percentage of people affiliated with an organization, whose main aim is research, is 15%. These are the people working with CENTA in the county.

Therefore it can be said that there is little research carried out in the area to adapt the promoted practices to the local conditions and production system or to even develop new conservation techniques. "Farmer experimentation" can be a helpful method though to stimulate the farmers' innovative potential, but it only works relatively localized. It needs to be complemented by more formalized research to make results comparable on a larger scale.

The investigated institutions seem to put most of their efforts in the promotion of practices. These are the standard conservation techniques found all over Central America (also see Chapter 2). These practices have all proven their usefulness to mitigate the erosion problem, but as demonstrated in Chapters 2 and 3 quite a number of them are not congruent with small-scale farmers circumstances, resulting in low adoption rates. Here the question arises if there is not more research in different forms needed to encourage an overall change in technology use towards more resource conserving practices.

But it seems to be difficult for the investigated institutions to use research to learn more about the suitability of the promoted conservation techniques for their target group. Hence, what other kind of mechanisms could the organizations employ to get feedback on their selection of practices? And how can this feedback be incorporated into the organization's work?

Most of the investigated institutions have some kind of feedback mechanism in place that helps them to adjust their work according to the response of their target group with respect to the promoted technolgy options (Figure 5-8). Two of the NARS, three of the NGOs and five of the "Other Organizations" work with informal surveys, internal and external evaluation commissions and farmers' workshops to evaluate the organization's work and the results of farmers' experiments.

Figure 5-8: Number of investigated organizations having some kind of evaluation mechanism for their work with soil conservation techniques

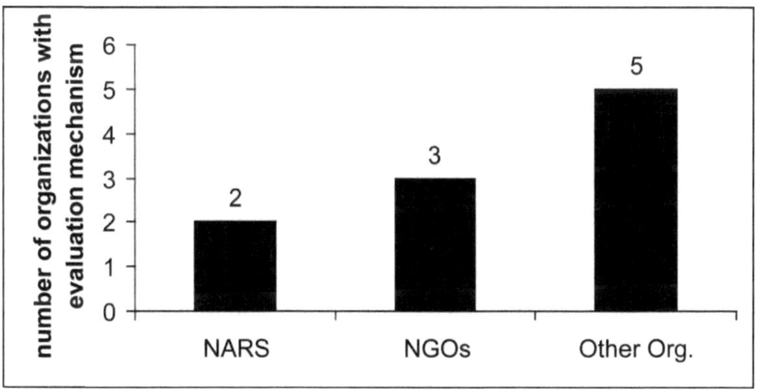

Source: own investigation, N = 14

Figure 5-9: Number of investigated organizations implementing changes in their work after the evaluation

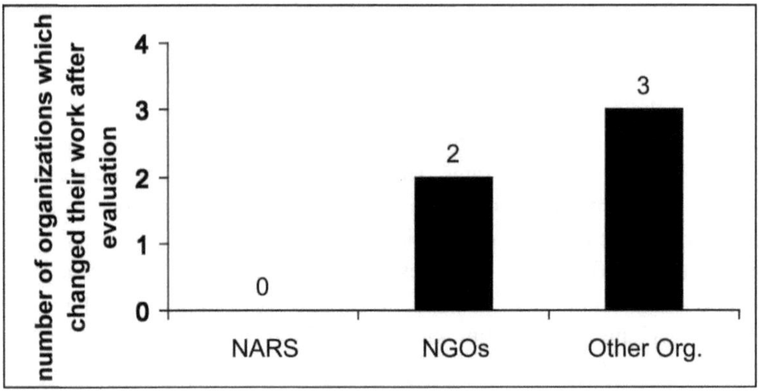

Source: own investigation , N= 14

Asked if and what kind of changes were made in the work of the institutions after these evaluations, it was surprising to see that not all organization representatives could answer this question or name any particular results of the feedback. This was especially the case with the NARS, but also with some of the "Other Organizations" and one of the NGOs (Figure 5-9). It seems that in these cases the feedback information did not reach the people it was intended for or there were no changes necessary, which would be a bit surprising.

5.4.3 The promotion of soil conservation techniques

The investigated organizations use a variety of different extension methods for their work with the farmers in the county and to promote the different soil conservation techniques (Table 5-8). The NARS rely here on the more traditional methods of inviting farmers to meetings in which new techniques are presented, visits of the technical staff to the farms and the organization of field days. Only CENTA, the public research agency of El Salvador, builds farmers groups with innovative farmers as the leaders of a community group. The same methods are utilized by NGOs and the "Other Organizations" as well, but they additionally employ a variety of other extension methodologies. While NGOs also put an emphasis on farmer meetings, field days and demonstration plots, they additionally engage in building community groups and providing the interested farmers with technical knowledge through courses. Helping farmers to experiment with particular new techniques and broadening their knowledge base in educational centers and by facilitating the exchange of experiences can be seen as further means to increase the human capital base in local communities. The "Other Organizations" employ as well these kinds of methods that lead to creating a broad technical knowledge in the communities through courses, experience exchanges and farmers' groups. Especially one of the projects (IICA-Holanda/LADERAS) tries to reach a long-term impact, which can be felt beyond the end of the project's intervention, by facilitating the creation of so-called 'Committees for Sustainable Development'. These committees then organize the NRM work within their communities. The project helps for example with providing committee participants with the requested technical knowledge and by documenting their experiences through publications. The idea behind it is to strengthen the innovative potential of the communities by providing them with the organizational and knowledge skills to find solutions themselves to resource management conflicts. One of the programs (PASOLAC) does not directly work with farmers but facilitates the work of other organizations through workshops and publications for their technical staff and through implementing validation plots for different techniques.

Table 5-8: Extension methods used by investigated institutions for their work with soil conservation techniques in Nueva Concepción

Extension methods used by investigated organizations	**NARS (3 org.)**	**NGOs (5 org.)**	**Other Org. (6)**
• Meetings with presentations etc.	+++*	+++	+++
• Farm visits of technical staff	+	+	+
• Field days/Field trips with farmers	++	+++	+++
• Farmers visits of institution	+		
• Demonstration plots/practical work	+++	++	++++
• Participation in educational centers		+	
• Farmer to farmer/exchange of experiences		+	+
• Farmers' experimentation		+	
• Building of farmers' groups in communities	+	++	++
• Technical training courses		++	+++
• Publications			+
• Not involved in direct extension, technical help for org. in area			+

* + = signifies 1 institution of this type (i.e. if there are 3 NARS up to 3 + in this category can be given)
Source: own investigation, N = 14

In the past the use of direct incentives, given to farmers in the form of agricultural inputs, vegetative material and seeds, tools and financial help for implementing soil conservation techniques, has been a frequently used tool for many institutions to facilitate the adoption of conservation practices. Reasons given as a justification include the need to compensate farmers for foregone income, which can be reduced in the initial stages of the use of soil conservation techniques (e.g. yield depression in the first years or reduced arable land area), the need to reduce the risk of poor farmers in implementing a new agricultural practice, and the compensation for additional labor often associated with certain soil conservation techniques (Giger 1999). Today these incentives are often criticized as it has been found that their use does not automatically lead to a permanent adoption, which continues after the termination of the work of the promoting organization. In addition, there are examples in which the diffusion of conservation practices could be achieved without these incentives or even without any intervention of promoting institutions at all. The direct incentives might even lead to reduced adoption as farmers not participating in the corresponding programs might not want to use the promoted practices until they also receive the associated benefits (Giger 1999, Dominguez *et al.* 1997, Schrader, n.d.).

Figure 5-10: Number of investigated institutions in Nueva Concepción giving direct incentives to farmers for implementing soil conservation techniques in 1999

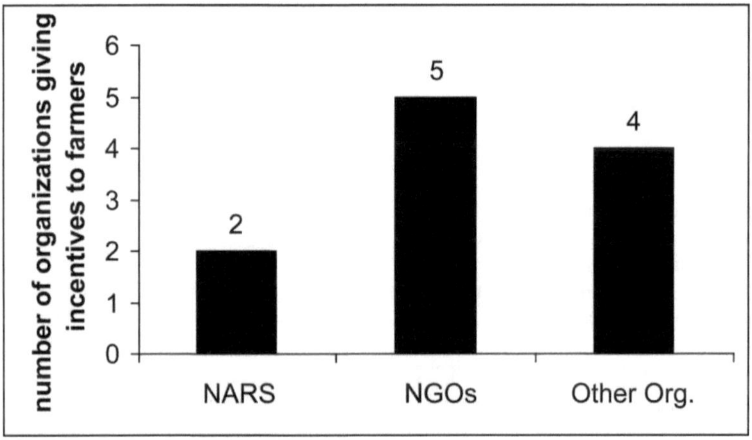

Source: own investigation, N = 14

Despite of these doubts, 11 of the 14 investigated institutions in Nueva Concepción still work with direct incentives for farmers (Figure 5-10). As can be seen in Table 5-9, two NARS organizations mainly provide farmers with vegetative material and agricultural inputs, like fertilizer or pesticides, to establish for example live barriers, tree nurseries or grow horticultural crops. Also the NGOs facilitate the introduction of conservation techniques with the respective seed and vegetative material and inputs. They also provide farmers with tools and give them specific credits for buying necessary inputs. In return these farmers have to produce vegetative material for other farmers. Also the "Other Organizations" work mainly with seed and vegetative material and agricultural inputs. One of them is additionally providing farmers with financial help and tools when implementing the conservation measures.

Table 5-9: Incentives for farmers given by the investigated organizations in Nueva Concepción for adopting the promoted soil conservation techniques

Type of incentive given to farmer	**NARS (3 org.)**	**NGOs (5 org.)**	**Other Org. (6)**
• Vegetative Material, Seeds	++	++++	+++
• Tools	+	+++	+
• Financial help/credits for implementation		+	+
• Agricultural Inputs	++	++	++

* + = signifies 1 institution of this type (i.e. if there are 3 NARS up to 3 + in this category can be given)
Source: own investigation, N =14

To finish the section on the promotion of soil conservation practices let us look at the main problems organizations encounter when working with these techniques in an area characterized by small-scale, subsistence farming (Table 5-10). These difficulties can be categorized either as problems due to structural factors of the area or as internal problems within the organizations. Almost all representatives of the surveyed organizations mentioned the difficulties they have in motivating farmers to fight erosion and conserve soil, often even if they are aware of the problem and face it on their farms. Reasons for this are manifold, but one important point is that farmers under subsistence conditions have to see a substantial impact on their income associated with the use of more soil conserving techniques before they might be willing to change their soil management practices. Therefore many organizations put an emphasis on the economic characteristics of conservation practices. Offering techniques without appropriate characteristics can lead to low adoption, observed by one of the

NARS and one of the NGO representatives. This motivation problem is closely related to a structural factor of the area, land tenancy (see also section 5.3).

Most small-scale farmers are land renters, for whom it does not make sense to invest in the improvement of land that is not their own. NGO representatives mentioned additionally that it seemed difficult for farmers to develop new ideas themselves. And one "Other Organization" experienced difficulties associated with the use of direct incentives.

There are also problems within the organizations. The “Other Organizations” seem to be more open to self-critique, as their representatives mentioned a number of problems. These range from missing new methodologies for improving the innovation process in the communities and following-up on the previous work done with farmers to coordination difficulties with other organization, internal planning or the availability of funding. NARS and NGO representatives hardly mention any of these problems.

Table 5-10: Main problems mentioned by representatives of investigated organizations encountered in their work with soil conservation techniques in Nueva Concepción

Main problems in work with soil conservation practices in Nueva Concepción	**NARS (3 org.)**	**NGOs (5 org.)**	**Other Org. (6)**
A. Problems with farmers			
• Difficult to motivate people to do something, even if they are aware of erosion problems	++	++	++++
• Difficult for farmers to develop new ideas themselves		++	
• Jealousy of farmers not included in org.'s work			+
• Low adoption of techniques, no incentives	+	+	
B. Structural factors			
• Land tenancy		++	+
C. Problems within organizations			
• Missing new methodologies to improve innovation process			+
• Missing systematic follow-up of farmers' training			+
• Planning of activities		+	+
• Availability of resources	+		+
• Co-ordination with other organizations			++

* + = signifies 1 institution of this type (i.e. if there are 3 NARS up to 3 + in this category can be given)
Source: own investigation, N = 14

In summary it can be said that the different types of organizations, which work in the promotion of soil conservation practices, differ in their overall range of activities and the kind of conservation techniques they offer. These differences can be associated with the technology selection process and its criteria, the degree of communication with farmers, and the information sources employed. In addition, not all organizations have work evaluations or information feedback loops in place that help to re-structure their work. Nevertheless, the range of soil conservation techniques offered to farmers does not vary greatly between the organizations; the variations result mainly from the emphasis that the organizations put on particular types of practices due to their selection criteria.

Another additional point should be emphasized here. There is little research going on in adapting the promoted practices to the area or developing new practices. Though some of the institutions encourage farmers to experiment on their farm, the results do not seem to be well documented or systematized. In addition, it seems to be difficult to motivate the innovative potential of farmers, as some of the interviewed representatives mentioned. If compared with the ideas postulated by the "Induced Innovation Theory" on the relationship between the need for research and technical change it is therefore questionable how technical change towards more resource conserving management practices can take place in this area.

5.5 Farmers' response to the offer of soil conservation techniques

Now that we have seen who offers soil conservation techniques and how this is done let us look at farmers' response.

In Table 5-11 some general characteristics of the interviewed farmers can be found. With an age of 49 years, the average age of farmers lies in the middle range. Their education level is in general relatively low as only 46% attended school at all and only 3 % went to a secondary school. With 70% of the surveyed farmers being landowners this percentage lies higher than expected compared with the above-mentioned figures on land ownership of the PROCHALATE and CENTA surveys (section 5.3.1). The reason for this can be the more stable political and institutional environment developing now after the end of the civil war. But it could also be an indicator that today more landless people leave the county in search of better opportunities. This might be especially the case with younger people, for which the higher average age of the interviewed farmers can be an additional indicator. The majority of land-owners (81%) only farm their own land and do not rent additional land. With 2.4 ha as the average farm size and also the average size of land owned by a farmer, these farms can be classified as subsistence farms that only produce enough to more or less cover the food requirements of the farm family. They do not generate much surplus for market sales. The average size of rented land at 1.5 ha is even smaller again indicating that most of the surveyed farmers were subsistence producers.

Table 5-11: General characteristics of surveyed farmers in Nueva Concepción

Characteristics	Level
• Average farmers age	49 years
• Visit of primary school	46 %
• Visit of secondary school	3 %
• Land owner	70 %
• Land owner who only farms on his own land	81 %
• Land owner who additionally farms on rented land	19 %
• Land renter (no own land at all)	30 %
• Average size of owned land	2.4 ha (3.5 mz)*
• Average size of rented land	1.5 ha (2.2 mz)*
• Average farm size	2.4 ha (3.5 mz)*

*1 mz (manzana) = 0.7 ha (hectar)
Source: own investigation, N = 76

Table 5-12 shows the main crops grown by the surveyed farmers. Maize is clearly the most important crop for these subsistence farmers as it is the most important staple in El Salvador. Interesting here is to note that it is grown more on rented than on owned land, which shows that most landless farm families merely live of what they can produce and thus mainly plant maize and beans. Beans are the second most important crop and also serve mainly for home consumption, while sorghum is often sold. Cash crops are rice and sesame, but also fruits and vegetables, which are only planted by a small number of farmers. Especially sesame seems to be an important cash crop for renters, as it is grown twice as much on rented land as on owned land. In the following we will investigate how these smallholders respond to the choice of soil conservation techniques they are offered. As we already know from the earlier mentioned "PROCHALATE" survey (section 5.3.1) the majority of farmers (75%) is aware of the erosion problem present in the area. But does this lead to a high adoption of conservation practices? Did the majority of farmers receive any of the incentives offered to facilitate adoption, either in form of direct incentives or in form of information or courses? And which are the practices used? Does the focus of the investigated institutions on different technology types match with the selection made by farmers?

To answer these questions we will first investigate which soil conservation practices are used by farmers in Nueva Concepción and how they obtained information about these techniques. Then the selection of practices farmers made will be compared with the focus of the investigated organizations on certain types of conservation practices.

Table 5-12: Crops grown by the surveyed farmers in Nueva Concepción

Crop	% of farmers growing crop on own land*	% of farmers growing crop on rented land**
• Maize	81	91
• Beans	59	52
• Sorghum	57	52
• Rice	8	0
• Sesame	6	12
• Fruit trees	6	0
• Yucca/Maniok	4	0
• Vegetables	4	0

* includes land owners that also rent land, figures only for their own land
** includes land renters, that also own land, figures only for rented land
Source: own investigation, N = 76

As a first observation (Table 5-13) it can be said that in the survey not one farmer was found who did not use at least one soil conservation practice on his fields. Often farmers would use several conservation practices in combination. This is already a promising result for the investigated organizations, showing that farmers are very well aware of the erosion problem and are searching for solutions. This is probably enhanced by the courses with information on the erosion problem and the implementation potential of different natural resource management practices that more than 50% of the surveyed farmers received. Few farmers (16%) obtained direct incentives in form of seed material, tools etc. This leads to the conclusion that also without direct incentives at least a few of the promoted conservation practices are attractive to farmers.

In Table 5-14 the percentages of farmers using one or more soil conservation techniques either on their own or on rented land are shown. First it can be seen that farmers only use practices promoted by the investigated organizations. Abolishing the burning of crop residues is by far the most popular technique and

Table 5-13: Characteristics of surveyed farmers in Nueva Concepción related to soil conservation techniques

Characteristics	% of farmers
• Adoption of at least one soil conservation practices	100
• Direct incentives received	16
• Information/Courses received	53

Source: own investigation, N = 76

is used by more than 90% of the surveyed farmers on rented and 85% on their own land. The so-called 'No-Quema' is one of the conservation practices most widely promoted in all of Central America, as the continuous burning of crop residues prevalent in many cropping systems today, has been identified as one of the main causes for soil erosion in the region. This practice is also considered to be a Productivity-Enhancing, Resource-Conserving technology as the effect of additional biomass on crop yields can often be seen within two to three cropping cycles.

The second most important conservation technique used in Nueva Concepción is constructing barriers made either from leguminous shrubs or grasses (live barriers) or from stones (dead barriers). 36% of landowners and 30% of land renters established live barriers. As this practice is relatively easy to implement and can serve a double purpose by retaining soil and providing organic matter, it is quite easy to understand why farmers use it relatively often. As stones are abundant on many fields, 38% of the farmers constructed dead barriers on their own land, while these physical structures, which do not have the same double function as live barriers, are not as popular with land renters (21%). Living fences, often with leguminous tree and shrub species, are established by 30% of the surveyed landowners instead of putting up a normal wire or wooden fence. For land renters this practice seems to be less attractive as only 15% implemented this technique.

Other changes in soil management practices, like contour tillage, contour ridges and cover crops, do not seem to attract farmers. In the case of cover crops this is especially surprising, as different legumes are used as cover crops by a large number of farmers in many parts of Central America under similar conditions. Thus either not the right type of legumes has been selected for Nueva Concepción or it is difficult for farmers to obtain the seed material. Another reason can be that cover crops have not been as strongly promoted in the area as other techniques. In the previous section we have seen that so far only a few organizations work with them. We will come to this point again later.

Apart from dead barriers, other physical structures seem to be of almost no interest to farmers, as only 8 % of landowners established drainage ditches and 2% individual terraces with fruit trees. Land renters do not use any of these practices.

Table 5-14: Adoption rate of soil conservation and agricultural practices promoted by the investigated organizations in Nueva Concepción

Promoted Soil Conservation and Agricultural Practices	**% of farmers using the technique on own land**	**% of farmers using the technique on rented land**
A. Changes in soil management practices		
• No-Burning practices with crop residue management	85	91
• Live barriers with grasses, *Gliricidia sepium*, *Cajanus cajan* or pineapple	36	30
• Contour tillage (curvas a nivel)	2	0
• Contour ridges (callejones)	0	0
• Legume intercropping/Cover crops	4	0
• Living fences	30	15
B. Physical Structures		
• Individual terraces	2	0
• Bench terraces	0	0
• Dead barriers of stone or straw	38	21
• Drainage ditches	8	0
C. Introduction of a complex system		
• Agro-forestry	0	0
• Tree planting/Reforestation	26	24
D. Practices related to natural resource conservation		
• Nurseries	0	0
• Land use layout design of the farm	0	0
• Protection/management of water wells	0	0
E. Agricultural Practices		
• Planting of fruit trees	2	0
• Improved pastures	0	0
• Crop diversification	0	0
• Pesticide management	0	0
• Use of organic fertilizers	2	0
• Horticultural crops	0	0
• Organic fertilizer pits	0	0

Source: own investigation, N = 76

Some of the complex land use systems are more appealing to farmers. Planting of trees is an important technique. It is usually done not as a real reforestation measure, but different timber or fruit trees are planted dispersed on the field. Growing annual crops is usually still possible. Thus it is not surprising that 26% of the landowners plant trees on their land. Interesting is that about the same percentage of land renters (24%) planted them on their rented fields. This is surprising as land renters usually do not have access to the land long enough to harvest the trees. Thus there either exist some long-term land rent agreements, which would be unusual for the area, or the landowners encourage their tenants to implement certain conservation measures. Agro-forestry is not mentioned by any of the farmers. As some of the tree planting practices described by the farmers seem to be similar to certain agro-forestry techniques it seems that farmers actually use these practices but do not call them by that name. Thus they probably do not see these techniques as part of a bigger system and are not aware of the interactions within the system.

The practices related to natural resource conservation were not mentioned by any of the farmers. This probably has something to do with the fact that these practices have only recently been promoted and are usually carried out more on community level. Thus not all farmers are involved in this kind of activity.

Farmers also mentioned only a few of the new agricultural practices. As in the survey farmers were asked to name all soil conservation techniques they apply on their fields they then did not mention these practices.

Thus if we ask which soil conservation technology type farmers prefer (Figure 5-11), we see that on own as well as on rented land farmers introduced mainly changes in soil management practices. More radical measures like implementing physical structures or complex land use systems play only a marginal role. Also the tendency of farmers to use physical structures more on own than on rented land can be observed. As these techniques require at the beginning a lot of labor and technical know-how and the time period until benefits can be received is longer than with changed soil management practices, long-term access to land has to be guaranteed for farmers before it makes economic sense for them to adopt these practices. This is also the case for dead barriers and the planting of trees, but some farmers mentioned that field owners sometimes encourage their tenants to implement or maintain these techniques. As mentioned before, general resource management practices are not used at all by the surveyed farmers.

How did farmers first learn about the soil conservation techniques they later adopted? In Table 5-15 the information sources of farmers are shown. Farmers received information from quite a number of different sources. For all adopted techniques NARS (and here mainly the national research and extension agency

Figure 5-11: Shares of the different technology types in total use of soil conservation techniques on own and on rented land in Nueva Concepción

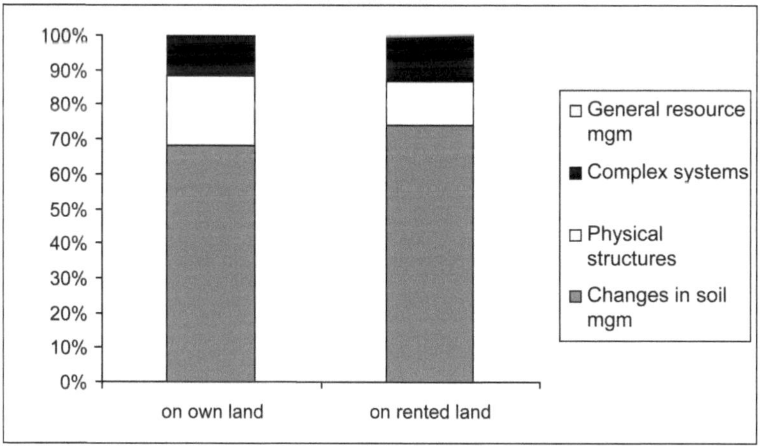

Source: own investigation, N = 76

CENTA) were the main information source. This has definitely something to do with the long period that the NARS exist and work in the area; all of them are active in the county for more than 10 years (Figure 5-3). Traditionally they used to be the only research and extension service in the area, whose task it was to do research about topics of major interest to farmers and deliver its results to them. Therefore they also played a significant role in providing farmers with new information and technologies. Nevertheless, usually only a part of the information or of the promoted practices were generated directly in the county. The transfer of information from other parts of El Salvador or even Central America was an important part of the work of the NARS. This is still the case today. The role the NARS play for many farmers makes it therefore even more important that they review carefully the kind of information they offer and, in the case of conservation techniques, on what kind of practices they focus.

Today the work of the NARS is complemented by a large number of NGOs and projects, as it is also the case in Nueva Concepción. But nevertheless, the second most important information source for the surveyed farmers are other farmers that already implemented one or more techniques successfully on their fields. This stresses the importance of visible demonstrations where farmers can evaluate a certain practice under circumstances similar to their own.

Table 5-15: Information sources for surveyed farmers in Nueva Concepción about the adopted soil conservation techniques

Promoted Soil Conservation Practices	% of farmers, who received information about a particular soil conservation technique from						
	NARS	NGOs	Other Organizations	Additional Organizations*	Own experience	Other farmers	Owner of field
A. Changes in soil management practices							
• No-Burning practices with crop residue management	47	3	11	4	8	17	8
• Live barriers with grasses , *Gliricidia sepium*, *Cajanus cajan* or pineapple	18	1	7	3	1	9	3
• Contour tillage (curvas a nivel)	0	0	1	0	0	0	0
• Contour ridges (callejones)	0	0	0	1	0	0	0
• Legume intercropping/Cover crops	0	0	1	0	0	0	0
• Living fences	13	0	3	0	0	5	1
B. Physical structures							
• Individual terraces	0	0	0	1	0	0	0
• Bench terraces	0	0	0	0	0	0	0
• Dead barriers of stone or straw	24	1	4	3	0	5	3
• Drainage ditches	3	0	0	0	0	1	0
C. Introduction of a complex system							
• Agro-forestry	0	0	0	0	0	0	0
• Tree planting/Reforestation	13	3	3	4	3	7	3
D. Practices related to natural resource conservation							
• Nurseries	0	0	0	0	0	0	0
• Land use layout design of the farm	0	0	0	0	0	0	0
• Protection/management of water wells	0	0	0	0	0	0	0

*Additional organizations = Two organizations that do not work in the area any more or whose representatives could not be interviewed.

Source: own investigation, N = 76

Despite the relatively short time that the "Other Organizations" are present in the area, they have already reached quite a number of farmers and are the third most important information source. Here especially the concept of community committees, which organize the conservation work within their village, seemed to have worked as farmers often named the 'Comites de Dessarrollo Sostenible', started by the IICAHolanda/Laderas Project, as their information source.

NGOs on the other hand were not named as often by the surveyed farmers. Reason for this is probably that most NGOs do not work in all districts of the county and many of them are only active in the area since a relatively short time period.

Farmers also often rely on their own experiences or, in the case of land renters, on the knowledge of the people owning the land they farm. Landowners especially encourage the use of no-burn practices and the planting of trees, which explains the high percentage of land renters using these techniques. In addition, there are some other institutions (called "Additional Organizations" in Table 5-15), such as Church organizations, through which farmers received some information on certain soil conservation techniques.

Figure 5-12: Shares of the different technology types in promotion by the investigated organizational types and use by farmers in Nueva Concepción

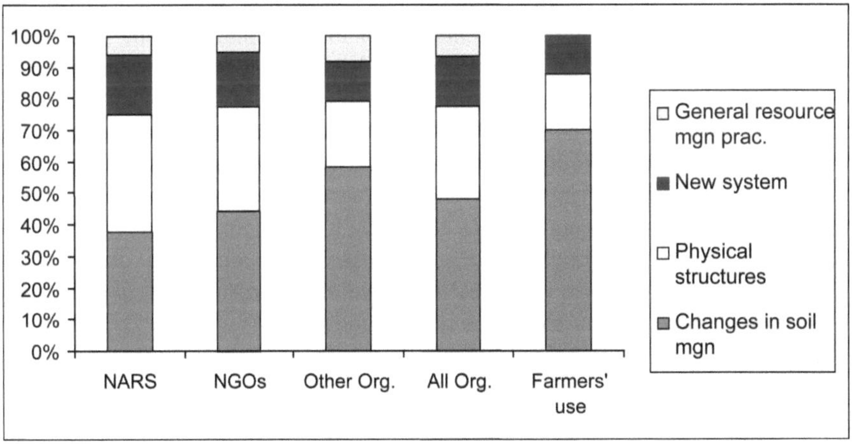

Source: own investigation

In summary it can be said that farmers use quite a number of information sources. The investigated institutions play a vital role in farmers' search for information. Therefore it is important that they select the information and the kind of practices they promote carefully. Figure 5-12 shows the differences in emphasis that the organizations put on the promotion of the three categories of soil conservation techniques and on General Resource Management practices. If the offer of all organizations together is compared with the farmers' response it can be seen that farmers mainly adopt Changed Soil Management practices. These practices are an important part in the overall offer of the institutions, but there exists also quite an emphasis on physical structures. As mentioned before (Figure 5-6), this tendency is strongest in the NARS, followed by the NGOs, while the "Other Organizations" are the strongest promoters of Changed Soil Management techniques. The NARS also put the strongest emphasis on promoting more complex systems, like agro-forestry.

Farmers mainly plant trees on their fields, but in general seldom use the complete set of practices belonging to the complex systems. General Resource Management practices are promoted the strongest by the "Other Organizations". As these techniques are utilized more on a community basis individual farmers do not mention them as part of their soil protection strategies. From this comparison the conclusion can be drawn that some of the institutions might need to re-structure their offer of conservation practices and the focus they put on certain conservation practice categories. If they want to foster change towards the adoption of more sustainable cropping practices, addressing farmers' demand for practices that stop erosion but do not directly imply large investments, specific technical knowledge or complete changes in the cropping system is crucial. Farmers' first choice in this respect seem to be Changes in Soil Management, which are in many cases quite compatible with subsistence farmers' circumstances in Central America (Chapter 2). Institutions therefore might want to revise the criteria for selecting which conservation practices to research or promote. In addition, improving information on agronomic and economic characteristics of practices can contribute to achieving higher adoption rates of offered techniques. Institutions might have good reasons though when deciding to continue promoting certain practices that have shown difficulties in adoption. Nevertheless, these reasons should be revised continuously.

5.6 Lessons learned

What are lessons learned about how soil conservation practices are supplied to farmers and farmers' response? And did the supply match what farmers were looking for?

First, we have seen that in Nueva Conception three institutional types of organizations offer soil conservation and general resource management practices to farmers: NARS, NGOs and "Other Organizations", which are present in form of projects and programs and are either funded by national or international agencies. NARS are the oldest institutions in the area while the two other organizational types mainly work in the county since the end of the Civil War.

The organizations had different reasons for selecting the county. For NARS and especially for the "Other Organizations" the situation of agricultural production and even more the natural resource management problems of the area were important. But while NARS have to offer technical support for the agricultural sector in general, the "Other Organizations" put special emphasis on small-scale farmers. In contrast NGOs came to the area more because of community related issues like, credit or land transfer programs or as part of the peace agreements after the El Salvadorian Civil War. Thus resource conservation issues are important for them as they affect the clientele of people they work with, but they are not their only or main focus. They cover a relatively wide range of activities, in which social, educational and human capacity building issues play a major role. Also for the "Other Organizations" human capacity and organization building and education is an important aspect of their work, but this is always connected with a special emphasis on resource conservation related issues. NARS on the other hand focus more on agricultural production issues, to which the proper management of the natural resource base is directly related.

All institutions use interactive, participatory methods to elicit the most pressing problems of their target groups. These are usually complemented by surveys, workshops with key informants or literature reviews. It is interesting to note that only NGOs mention that local communities also pursued them for help.

As solutions to the natural resource degradation problem of the area the investigated organizations offer a wide range of EIs. The offered techniques can be grouped into five main categories: Changes in farmers' soil management, physical structures, complex resource use systems, general practices related to natural resource management and certain agricultural practices. All institutional types work with conservation techniques of each category and thus offer a similar range of practices. But they put a different emphasis on particular technology types. NARS concentrate more on physical structures and changes in soil management. NGOs focus less on physical structures and the other technology types become more important. With the "Other Organizations" this tendency becomes even stronger as they put an emphasis on changes in soil management. But they still also promote selected techniques from the other categories.

Reasons for the differences in emphasis can be found in the selection criteria of the organizations for choosing the techniques to work with. While NARS mainly try to be responsive to discussions with communities, the other two organizational types have more specific selection criteria. Especially in the case of the "Other Organizations" these criteria include next to farmers' suggestions agronomic and in particular economic characteristics of the techniques. Also NGO representatives mention quite a number of important economic criteria. Thus the representatives of these two institutional types seem to be well aware of the importance that economic implications of implementing conservation techniques have on adoption. This observation does not confirm the first sub-hypothesis about the lack of consideration of short- and medium-term profitability of the promoted techniques by the investigated organizations. The "Other Organizations" seem to have reacted to this observation the strongest and adjusted their technology range to include mainly changed soil management techniques, which are in their majority relatively easy to implement, and additionally show productivity enhancing effects on farmers' fields in a relatively short period.

The differences in selection criteria and the resulting choice of soil conservation practices is influenced by the people involved in the final decision on offered techniques and by the information sources used by the different organizations. In NARS and the "Other Organizations" farmers have a stronger say in the final decision than in NGOs, which stated that their technical personnel mainly takes this decision. But in NGOs as well as in the "Other Organizations" the staff seems to be more encouraged to search for information, which has contributed to the wider range of selection criteria. Nevertheless, in all organizations also headquarter training plays an important role.

Research on the promoted techniques is only carried out by a small number of institutions. It is all adaptive research and, in the case of NGOs and the "Other Organizations", farmer experimentation. This research helps to adjust the offered practices to local circumstances, but the question remains how the results of farmer experimentation can be compared with each other and what effective dissemination mechanisms can be found. This confirms the second sub-hypothesis about the lack of research on the promoted practices. Organizations put the major part of their efforts on the dissemination of techniques researched in other parts of El Salvador or Central America. This transfer of information and technologies is important, but it needs to be combined with research in different forms to supply well-adapted practices to farmers. The overall low research capacity in the area can be seen as an important weakness in reaching a transition to a more sustainable use of natural resources. As laid out in the "Induced Innovation Theory", an active exchange between technology suppliers and users is needed to develop practices that answer to resource constraints of farmers. With growing degradation techniques will become more important that conserve the agricultural production potential and farmers will start to search for these practices. The communication base between the organizations working in the promotion of these techniques is being established in the moment. But it needs to be complemented by mechanisms that help to bring new technology options to farmers.

In addition to research, which helps to determine the suitability of their technology choice for farmers, organizations also need to put certain feedback mechanisms in place to evaluate the effectiveness of their work. Most investigated institutions mention different systems in form of surveys, evaluation commissions, etc., but a much smaller number of organizations actually implemented changes according to these evaluations. This indicates that the feedback loop within the organizations is often not closed.

All organizations realize the importance of demonstrations and field days to show farmers the effectiveness of the proposed conservation practices. NGOs and the "Other Organizations" furthermore stress the transfer and generation of know-how and technical understanding for farmers to increase the knowledge base on Natural Resource Management (NRM) at the community level. Also direct incentives in form of seed and vegetative material, tools or access to credits and agricultural inputs are used to facilitate adoption. Doubts have been raised though in recent literature about the effectiveness of these measures. These incentives did not seem to have influenced adoption behavior very much in Nueva Concepción. Nevertheless more in-depth research would be needed to clarify this point.

The main problems the investigated institutions encounter in their work in the county are either related to the direct work with farmers or are more inner-institutional problems. Structural factors hindering adoption are seldom mentioned. Motivating farmers to really implement changes in their own behavior seems to be the main difficulty.

Farmers in Nueva Concepción are very much aware of the soil degradation problem in the county. All of the surveyed farmers implemented at least one soil conserving practice on their own or on rented fields. The majority of the adopted EIs belong to the category of changes in soil management. Among these are the two most important PERC practices promoted in Central America: no-burn systems and leguminous cover crops. This adoption behavior of farmers justifies the emphasis different organizations put on this type of conservation technologies. But organizations also put a considerable part of their research and promotion efforts into other kinds of practices. This confirms the postulated general hypothesis that the offer of soil conservation practices only partly matches those demanded by farmers. It also shows that for farmers cost-benefit considerations and the short-term implications of the use of these practices are important, as one common characteristic of techniques in this category is that they show positive effects on soil fertility faster than other technology types. Thus looking at characteristics of conservation techniques in general and on their economic particularities in particular can be a useful strategy of technology suppliers when making their selection. Some of the investigated organizations, in particular NARS, will have to re-structure their technology selection criteria. Asked about their information sources for the adopted practices, farmers mainly named the NARS (and here mostly the El Salvadorian public research and extension service CENTA) as their most important source, followed by other farmers and the "Other Organizations". This is not surprising as traditionally the NARS are the main partners for farmers through the public extension service. The other institutions present in the area today though are beginning to play an important complementary role to NARS. Good co-operation and the exchange of information and experiences between all the organizations working in the field of soil conservation is thus of vital importance, as it helps to disseminate information through channels known to farmers or opens up new ways. These can for example lie in the closer co-operation with farmers in the field of research and setting research priorities. Setting up community groups, what a number of the investigated organizations, including NARS, have started, can be one of these new possibilities.

6 Summary and Conclusions

In recent decades the awareness of the progressing degradation of natural resources and the importance of their long-term, careful use has increased worldwide. Confronted with growing degradation problems, such as the loss of biodiversity, air, water and soil resources, sustainable development is seen today as the only possible solution to guarantee the survival of human and natural systems in the future. Sustainable development and resource use combines elements of an ecologically sound utilization of natural systems with the notion of economic and social development. In the last few years discussions have centered on how to put the sustainability concept into practice. Implementation of sustainability plans face a number of difficulties, including the complexity of the concept and the fact, that successful solutions to fight resource degradation are mostly location specific, impeding the transfer of proven technical solutions without thorough adaptation. One important issue that must be understood, if we are to overcome these problems, is the role of technical change, especially in the agricultural sector.

For agriculture, as the most important and demanding form of land use worldwide, the careful management of fragile soil resources is the very basis for its long-term success. Mismanagement of these indispensable resources and the neglect of rules based on a long-term vision for its use are common in many places. Wind and water erosion are the most prevalent features of this mismanagement, leading to a constant threat to soil fertility and, with this, agricultural production. Central America can serve here as a clear example, as, in particular, water erosion is one of the most threatening problems for agriculture in the region. It has been estimated that up to 80% of the agriculturally used land in Central America is affected by human-induced soil degradation (ISRIC/UNEP 1991). The main reasons for this high rate of soil problems are the agro-ecological characteristics and the land use patterns prevalent in the area. The hilly topography and the long rainy season require careful management of soil resources to avoid the loss of topsoil and to prevent land slides. In addition, many marginal, especially hillside areas, are farmed by small-scale producers, whose land management practices are guided by rather short-term production goals to guarantee the survival of their families. In many areas the more productive farmland is concentrated in the hands of a relatively small number of large farmers, who favor land-intensive cattle production as their main activity. This skewed distribution of land ownership can be seen as the long-term result of polices favoring the production of cash crops for export in combination with insecure land property rights, which pushed the majority of small-scale farmers step by step into marginal and agricultural frontier areas. This development, combined with a growing population, has given momentum to the conversion of forestland to farmland used by smallholders. Once soil productivity for crop production decreased after a few years and as small-scale farmers did not have the capacity to invest in maintaining soil fertility, the land was taken over by larger farmers, mainly cattle producers, and converted into pastures.

Farming practices and soil conservation techniques have been developed by scientists together with farmers over the last decades to control and minimize erosion under different agro-ecological conditions. These techniques include practices to build physical structures, change soil management practices or introduce complex land use systems. Also in Central America these techniques were and are promoted by a large number of governmental and non-governmental organizations. But despite of this progress in knowledge and technical abilities the adoption of these practices is low. This is particularly the case for farmers in developing countries and also holds true for Central America. These low adoption levels remain a lasting and pressing concern in the region. Here the question arises, how technical change aimed at the protection and sustainable use of natural resources can take place under subsistence farming conditions. More specific, we can ask for the important factors influencing the decision-making process of farmers concerning soil conservation technologies. In addition, the question needs to be investigated what are the policy and institutional implications concerning the system for generating environmentally sound soil management practices in Central America.

A first step to answer these questions is to have a closer look at the differences between agricultural techniques geared primarily towards increasing farm profitability, i.e. Commercial Innovations (CI), and practices that are primarily directed towards improving land management, i.e. Environmental Innovations (EI). CIs were developed with the main objective of increasing farm productivity, and with this, farmers' income by either boosting agricultural output in a relatively short period of time or reducing total input costs. It is thought that all farmers under similar circumstances can use them. CIs provide proven economic benefits to farmers and if farmers behave according to a profit-maximizing rationale, adoption will take place in a relatively short period. EIs are developed to maintain a functioning agro-ecological system. In general their implementation shows economic benefits to farmers only in the long run. While the farmer has to bear implementation costs, he does not capture all of the benefits from the use of these practices; society as a whole also benefits by avoiding resource degradation. EI implementation implies the understanding of the complex linkages between soil, water, plants, and livestock, and their utilization can be quite labor or capital intensive. Under subsistence conditions these characteristics of EIs make their adoption difficult. The substantial differences between the two technology categories result in the need to employ different policy instruments and extension approaches and to re-think current technology development strategies. For CIs it might be only necessary to adjust external farm circumstances to facilitate adoption, with EIs this kind of adjustment is helpful but not sufficient.

A number of theories have tried to explain the role and direction of technical change and the factors governing farmers' behavior when searching for new technologies. The first theory to incorporate technical change as an endogenous variable in the process was the "Induced Innovation Theory", elaborated by Hayami and Ruttan (1985). Today this theory has been developed further and new theories have emerged, but the "Induced Innovation Theory" still serves as a good basis to explain technical change under various natural and socio-economic conditions. In this context farmers' search for possibilities to replace a relatively scarce, and therefore more expensive production factor with a relatively abundant and therefore cheaper one, is seen as the fundamental mechanism. This induces public and private technology suppliers to provide a specific set of technological options. Technical change is also seen as influencing the institutional settings in which farmers make their adoption decisions. Thus the theory combines different mechanisms working at various scales to explain the adoption of new technologies.

Most research carried out so far on testing this theory under different institutional and cultural settings focused on the development and adoption of CIs. Within the framework of the theory EIs can be viewed from three different perspectives: 1) EIs treated equally to CIs, 2) short- versus long-term profitability aspects, 3) 'Productive capacity of natural resources' as a separate production factor.

If the adoption of an EI is analyzed equally to a CI, then it will have to be investigated what this EI is able to contribute to saving one of the three production factors land, labor, or capital. Different EIs can then be classified as more labor or capital intensive per unit of land. But here a fundamental aspect of EIs that is not considered, namely their purpose of establishing or maintaining functioning agro-ecological interactions and thus contributing to the reduction of environmental degradation.

Investigating both technology groups from the perspective of short- versus long-term profitability, demonstrates that in general CIs are profitable in a relatively short period of time, i.e. within one cropping season. EIs become profitable only in the long run, i.e. after more than one cropping season. This reduces the likelihood that farmers and in particular small-scale farmers perceive most EIs as being more profitable as their traditional farming system. This specific characteristic of EIs places them in the priority list of farmers behind CIs. There exist two ways for this preference to be changed: First, EIs include in addition to their resource conservation aspect an equally important productivity enhancing component. These practices can then be viewed as 'Productivity-enhancing, resource-conserving' (PERC) or 'Overlap' Technologies (Vosti and Reardon 1997). They are designed to address the sustainability issue together with the need to sustain economic growth through productivity increases. A PERC or Overlap Technology can be viewed either as an EI with a productivity increasing aspect or as a CI that has some degradation reducing properties. Which one of the two aspects is regarded as the most important depends on the perspective of the technology user. For this class of techniques the inducement mechanism of the "Induced Innovation Theory" works in the same way as for normal CIs, as they are able to address the short-term profitability demands of farmers. Different practices have been identified as PERC or Overlap Technologies, which are in Central America legume cover crops and no-burn systems with crop residual management. Another option to change farmers' ranking with regard to CIs and EIs is, if farmers begin to place a value on the importance of well functioning agro-ecological interactions by understanding the relationship between these interactions and farm productivity. This might lead farmers to forgo profits today in favor of larger profits in the future. Under subsistence conditions though, farmers may not be in a position to lose any gain in the current period since this may jeopardize their survival.

A third way of interpreting EI adoption within the "Induced Innovation Theory" is to go back to the original reason for EI development, namely the maintenance or establishing of functioning agro-ecological interactions. Here the question arises whether these interactions can be interpreted as a production factor similar to the neo-classical factors of labor, capital, and land, which jointly all define the production function of a particular agricultural system. Such a factor can be named the 'Productive capacity of natural resources' in general and of soil resources in particular for the case of soil conservation practices. Similar to the other production factors, an increase in the quantity or quality of the input of the production factor 'Productive capacity' results in higher output. Under such an interpretation the profitability of the farmers' use of inputs becomes, at least in the long run, explicitly dependent on the production factor 'Productive capacity of soil resources'. If this factor deteriorates due to the occurrence of environmental degradation, and becomes scarcer, technologies that save on this factor will have a higher potential for adoption by farmers than other technology options. For its practical implementation though, this production factor has a decisive problem. It is difficult to assign a monetary value to the 'Productive capacity', making it difficult to compare its importance to the farmer with the importance of other production factors. The price mechanism, which signals farmers' resource scarcities, thus may not work here. Hence, even if a farmer faces environmental degradation on his farm, it is difficult for him to assess what this really means for him in terms of a decline in income.

The difficulties in assessing the importance of 'Productive capacity' can result in problems for technology developers to estimate the farmers' need and interest in implementing a technology with explicit EI characteristics. Therefore farmers' and technology suppliers' perceptions can differ substantially, misleading scientists to develop practices that do not correspond with the resource constraints as perceived by farmers.

For investigating some of the aspects of technical change with respect to EIs, two case studies were carried out in Central America. The main focus of the first case study was on factors influencing the adoption of a PERC technology by smallholders in Guatemala. The second case study dealt with the organizations and their methods of selecting and promoting soil conservation technologies in El Salvador. These two case studies examined the usefulness of the concepts developed by the "Induced Innovation Theory" for EI adoption and development.

The first case study was carried out in the Polochic Valley, Guatemala, and investigated the existing maize production system, the knowledge and perception of small-scale farmers of soil erosion, their use of soil conservation technologies, the adoption of Mucuna as a cover crop in maize, and the interactions between economic, social and institutional factors that lead to the use, non-use or abandonment of Mucuna. The Polochic Valley is an agricultural frontier area, where soil erosion is a visible problem and where farmer reports of declining in yields, especially in hillside areas, are frequent. One of the main reasons is the continuing use of soil degrading farm management practices, like burning of crop residues in hillside areas, the shortening of fallow periods and the increasing number of fields that are cropped continuously. Maize is the main crop grown by small holders in the valley and is planted in both growing seasons. Traditionally farmers tried to rotate among their fields but increased pressure on arable land drove many of them to abandon this system.

The herbaceous legume Mucuna has been used in the valley since the 1930's and is today widely known and intercropped with second season maize. The legume provides a thorough soil cover and is known to boost maize yields substantially. Nevertheless, its long growing cycle allows farmers to use their fields only once a year, in the second growing season, while the field lies under

Mucuna fallow during the following dry season and the next years' first season. This conflicts with farmers' need to intensify their cropping system and farm their fields twice a year. On the other hand the long-term soil improving characteristics of the Mucuna system allow the abandonment of long fallow periods necessary in slash-and-burn systems to restore soil fertility. Here it becomes evident that results with respect to the usefulness of the system for the farmer are quite different depending on if this system is analyzed from a short- or long-term perspective. From the short-term perspective the utilization of Mucuna prevents the more intensive use of available land resources. Therefore, if land becomes scarcer, technical change will take a path that saves land. The likely consequence is a stepwise abandonment of the Maize-Mucuna system. Case study results show that at present this development is taking place in the valley, demonstrating that the majority of farmers possess a rather myopic perspective. In contrast, if a long-term perspective is chosen, the Maize-Mucuna system constitutes a land use intensification. The Maize-Mucuna-System allows the continuous cropping of fields without long-term fallow periods, but without a serious decline in soil fertility. In this case it is likely that maintaining soil fertility in the long run is a decisive factor guiding the choices of land users and therefore directing technical change. Since at present linkages between farmers and institutions promoting a shift from a short to a long-term perspective are weak, this urgently required development is unlikely to occur.

A county in El Salvador was chosen for the second case study. The focus of this study was on the concepts, selection, and performance of organizations working in the development and promotion of soil conservation techniques. Land scarcity and soil erosion are two of the major problems in the area, where smallholders mainly grow maize, beans and some cash crops. Soil conservation practices are promoted by three different types of governmental and non-governmental organizations, which are the NARS, NGOs and "Other organizations" (development projects and programs). All institutions use interactive, participatory methods to elicit the most pressing problems of the target groups. With regard to solutions to fight the erosion problem, which are well known to farmers, the organizations offer a wide range of techniques. These practices can be grouped in five main categories: 1) changes in farmers' soil management, 2) physical structures, 3) complex land use systems, 4) general resource management practices, and 5) specific agricultural practices. All institutions work with practices from each category, but each puts a different emphasis on particular technology types. NARS concentrate on physical structures and some changed soil management techniques, NGOs focus less on physical structures, while this tendency becomes even stronger with the "Other Organizations". These mainly promoted changed soil management practices, which are easy to implement and often show productivity enhancing effects in a relatively short period of time. Reasons for the differences in focus can be found in the process through which the institutions select the practices to promote. NARS try to follow results of community meetings and farmers' suggestions, while the two other organizational types put in addition more emphasis on understanding the agronomic and in particular economic characteristics of the offered techniques. They seem to be well aware of the importance of the economic implications of implementing conservation techniques for farmers under subsistence conditions.

Little of the research necessary to adapt the promoted conservation practices, or develop new ones, is carried out in the area. As the "Induced Innovation Theory" states, an active exchange between technology suppliers and users is needed to develop practices that answer to the resource constraints of farmers and therefore provide the basis of technical change. The investigated institutions are mainly relying on technologies that were researched and developed outside the area. The transfer of research results and information on soil conservation practices is important, but insufficient to fully compensate for the missing guidance based on adaptive research carried out under local conditions. In addition, it seems that internal feedback mechanisms to evaluate the effectiveness of the work carried out by an organization need to be improved. Only a few mention that changes were introduced after work evaluations took place. This is an additional constraint to developing and promoting location specific solutions to the degradation problem.

Investigating farmers' response to the offer of soil conservation practices, it becomes clear that farmers are not only aware of the degradation problem but are trying to find solutions. All interviewed farmers implemented at least one conservation practice on their own or on rented fields. Abandoning the burning of crop residues together with residue management practices are by far the most popular techniques. Also other practices from the category of changed soil management practices are preferred by farmers. This adoption behavior seems to justify the emphasis that different institutions place on these types of conservation technologies. Their behavior also shows that for farmers, cost-benefit considerations and the short-term implications of the use of these techniques are quite important. This leads to the conclusion that it is very important for technology suppliers to make sure that the promoted technologies match farmers' location-specific economic and agronomic conditions.

A number of conclusions regarding the elements that govern the direction of technical change towards increased sustainability in land use systems in general and conservation of soil resources in particular can be drawn.

- Technology suppliers and promoting institutions must realize and accept that substantial differences exist between commercial and environmental innovations. Therefore they need to adjust their policy instruments and extension approaches accordingly.
- The short- versus long-term profitability aspects, which are characteristic for commercial and environmental innovations respectively, reduce the likelihood farmers, and in particular subsistence farmers, perceive environmental innovations as being more profitable for them than their traditional farming practices. Here specific efforts are required to change farmers' preferences by adjusting farmers' knowledge base and other farm conditions.
- Due to their short-term productivity-enhancing effects, PERC or Overlap technologies reduce the disadvantage that environmental innovations have in the farmers' perspective. Consequently they should receive a high priority on the list of promoted soil conservation technologies by technology suppliers. In addition, the technology development process must be geared towards this combination of economic and ecological aspects and both characteristics of technologies must be investigated according to location-specific requirements.
- Legume intercropping systems emerge as a promising PERC technology. Since they are easy to implement and provides short-term economic benefits together with erosion protection, their further development and adjustment should receive high priority in soil conservation development and promotion.
- Since the soil conservation techniques offered to farmers only partly match with farmers' demand, the technology selection process inside the institutions involved must be improved and additional adaptive research should be undertaken to tailor the selected techniques more precisely to particular farmers' conditions. Without continuous investments in researching and developing new natural resource management techniques, which also have a strong productivity enhancing component, it is not very likely that adoption levels of soil conserving techniques increase in Central America. This will then have serious consequences not only for the natural resource base of the region, but also increase the pressure on small-scale farmers' livelihoods.
- A closer co-operation among organizations involved in promoting soil conservation technologies with farmers in the fields of research and research priority setting as well as the formation of community groups will provide opportunities to improve adoption rates at the farm level.
- A number of mechanisms laid out in the "Induced Innovation Theory" can be helpful to explain farmers' behavior and the constraints the adoption of soil conservation technologies is facing under small-scale farmers' conditions. But the theoretical framework should be extended to accommodate the fact

that it is difficult to apply a common value to the factor describing the 'productive capacity' of natural resources. This is necessary to employ the price mechanisms, which guides the direction of technical change and is described as a central element of the Theory. In addition, this extension will be the only way to increase awareness of external costs connected with resource degradation in farmers' and society's perspective.

Reaching the transition from a short-term, short-sided to a long-term, sustainable use of natural resources requires understanding the linkages between a number of natural, economic, social, and institutional conditions in their location-specific context. Smallholder agricultural systems in developing countries reveal particularly well what happens if these linkages are distorted and/or neglected by system users and beneficiaries. Overuse and resource degradation are logical and visible consequences. To reverse this trend the development of new, resource-conserving technologies is a decisive step. But these practices will only be successfully adopted by small-scale farmers if they address in addition farmers' most pressing resource constraints. Overcoming these limitations requires for technology users to broaden their knowledge base with regard to maintaining functioning agro-ecological systems. On the side of technology suppliers a tailoring of offered solutions to the location-specific needs of their clients is demanded. Both can only be reached by strengthening regional research capacities, by directing research priorities towards production technologies with a strong conservation component, and by further developing mechanisms which help to incorporate farmers' demand into the research agenda for natural resource management practices. In this way the gap between private costs incurred for resource conservation and benefits received by society from its implementation can be narrowed. Without these fundamental changes in natural resource management the necessary transition is unlikely to take place and a break down of regional natural and social system is likely to occur sooner or later.

References

AHMAD, S., 1966: On the Theory of Induced Invention. *Economic Journal* 76: 344 – 57

ALMENDARES, R. D., ESTRADA, J. V., LEONARD, D., 1995: *Análisis Beneficio/Costo de Sistemas de Producción Mejorados*. Proyecto LUPE, Tegucigalpa, Honduras

ANDERSON, J.R., THAMPAPILLAI, J., 1990: *Soil conservation in developing countries: Project and Policy implications*. Policy and Research Series, No. 8. The World Bank, Washington D.C., USA

ASHER, H. B., 1976: *Causal Modeling*. Series Quantitative Applications in the Social Science No. 07 - 003. Sage Publications. Beverly Hills/London

AWANG, D., BUCKLES, D., ARNASON, J. T., 1997: The photochemistry, toxicology and processing potential of the cover crop Velvetbean (cow(h)age, cowitch) (*Mucuna Adans. Spp.*, Fabaceae). Paper presented at the International Workshop on Green Manure-Cover Crop Systems for Smallholders in Tropical and Subtropical Regions, 6-12 Apr. Chapeco, Santa Catarina, Brazil

BALAND, J.-M., PALTTEAU, J.-P., 1996: *Halting the Degradation of Natural Resources*. FAO and Clarendon Press, Oxford, UK

BARBIER, E.B., 1988: *The Economics of Farm-Level Adoption of Soil Conservation Measures in the Uplands of Java*. Environment Department Working Paper No. 11. The World Bank, Washington D.C., USA

BARRETO, H., 1994: Evaluation and utilization of different mulches and cover crops for maize production in Central America, in: THURSTON, H. D., SMITH, M., ABAWI, G., KEARL, S.(eds.), *Tapado Slash/Mulch: How Farmers use it and what Researchers know about it*. CATIE and CIIFAD, Ithaca, USA

BATIE, S.S., 1986: Why Soil Erosion: A Social Science Perspective, in: LOVEJOY, S.B., NAPIER, T.L. (eds.), *Conserving Soil: Insight from Socio-Economic Research*. SCSA, pp. 3 - 14, Ankeny, USA

BIOT, Y., XI., L., 1993: Assessing the Severity of the Problem and the Urgency for Action, in: BAUM, E., WOLFF, P., ZÖBISCH, M.A., *Acceptance of Soil and Water Conservation. Strategies and Technologies*. DITSL, Witzenhausen, Germany

BINSWANGER, H., 1974: A Microeconomic Approach to Induced Innovation, *Economic Journal* 84: 940 – 58

BISHOP, J., ALLEN, J., 1989: *The On-site Cost of Soil Erosion in Mali.* Environment Working Paper No. 21. The World Bank, Washington D.C., USA

BLAIKIE, P., BROOKFIELD, H., 1987: Defining and Debating the Problem, in: BLAIKIE, P., BROOKEFIELD, H. (eds.), *Land Degradation and Society.* Methuen & Co. Publ., London, UK

BUCKLES, D., TRIOMPHE, B., SAIN, G. 1998. *Cover Crops in Hillside Agriculture: Farmer Innovation with Mucuna.* IDRC/CIMMYT. Mexico DF, Mexico

CARE Internacional, no date: *Diagnostico Participativo Comunitario, Jolomijix II-B, Panzós, Guatemala.* Unpublished manuscript.

CARSKY, R.J., TARAWALI, S.A., BECKER, M., CHIKOYE, D., TIAN, G., SANGINGA., N., 1998. *Mucuna – herbaceous cover legume with potential for multiple uses.* Resource and Crop Management Research Monograph 25, International Institute for Tropical Agriculture IITA, Ibadan, Nigeria

CENTA/MAG,IICA-Holanda/LADERAS, 1996: *Resumen-Diagnóstico Agrosocioeconómico del GyTT-02 Nueva Concepción Generación y Transferencia Tecnológica.* IICA; San Salvador, El Salvador

CHAVEZ, E. L., 1994: *Desarrollo de tecnología en maíz – Abonos verdes en el Polochic.* Informe de un proyecto de investigación. ICTA, Panzós, Guatemala

CHAVEZ, E. L., 1995: *Informe técnico 1994/95 – Desarrollo de tecnología maíz-abono verdes en el Polochic.* ICTA, Panzós, Guatemala

CLARK II, E.H., 1987: Soil Erosion: Off-site environmental effects, in: HARLIN, J.M., BERARDI, G.M. (eds.), *Agricultural Soil Loss: Processes, Policies and Prospects.* Westview Press, London, UK

COMISION CENTROAMERICANA DE AMBIENTE Y DESARROLLO, 1998: *Estado del ambiente y los recursos naturales en Centroamérica.* San Jose, Costa Rica

CUESTA, M.D., 1994: Economic Analysis of Soil Conservation Projects in Costa Rica, in: LUTZ, E., PAGIOLA, S., REICHE, C. (eds.), *Economic and Institutional Analyses of Soil Conservation Projects in Central America and the Caribbean.* World Bank Environment Paper No. 8, The World Bank, Washington, D.C., USA

DAVIDSON, R., MACKINNON, J., G., 1993: *Estimation and Interference in Econometrics.* Oxford University Press, New York, USA

DEGROOT, J.P., AMBROGUI, R., LOPES JIMENEZ, M., 1997: Production Systems in the Humid Tropics of Nicaragua: A Comparison of Two Colonization Areas, in: DEGROOT, J.P., RUEBEN, R. (eds.), *Sustainable Agriculture in Central America.* MacMillan Press Ltd., Houndsmills, UK

DOMINGUEZ, A., PEREZ, A., VENTURA, C., MERLO, H., MENDOZA, J., SCHRADER, K., CLARO, L., CONTRRERAS, V., 1997: *Estudio de adopción de tecnologías de agricultura sostenible en Guajiquiro y Opatoro, La Paz, Honduras.* APAS/INTERCOOPERATION/PASOLAC, Managua, Nicaragua

DVORAK, K.,A., 1996: *Catalogue of soil conservation practices and projects in Central America.* Hillsides Program. International Center for Tropical Agriculture, Internal Report, July 1996, Tegucigalpa, Honduras.

EL-SWAIFY, S.A., ARSYAD, S., KRISHNARAJAH, P., 1983: Soil Erosion by Water, in: CARPENTER, R.A. (ed.), *Natural Systems for Development. What Planners need to know.* EAPI, MacMillan Publ., New York, USA

EL-SWAIFY, S.A., 1994: State of the Art of Assessing Soil and Water Conservation Needs and Technologies, in: NAPIERI, T.,L., CAMBONI, S. M., EL-SWAIFY, S.A. (eds.), *Adopting Conservation on the Farm.* Soil and Water Conservation Society, Ankeny, USA

EILITTÄ, M., 1998: *On-Farm Assessment of Maize-Mucuna Systems in the Los Tuxtlas Region of Southeastern Veracruz, Mexico.* Ph.D. thesis, University of Florida, USA

ELLIS-JONES, J., SIMS, B.,1995: *An appraisal of soil conservation technologies on the hillside farms in Honduras, Mexico and Nicaragua.* Project Appraisal, Vol. 10, Issue. June 1995, pp. 125 - 132

ERENSTEIN, O., 1999: *The Economics of Soil Conservation in Developing Countries: The Case of Crop Residue Mulching.* Ph.D thesis, Wageningen University, Wageningen, The Netherlands

FOURNIER FAUCHÈRE, J. M., unpublished: *Propiedades del Suelo y Rendimiento del Grano, Producción de Rastrojo y Biomasa a la Cosecha en 29 Ensayos de la Cuenca del Polochic, Guatemala, Postrera 1996.* CIMMYT, Guatemala CD, Guatemala

FOURNIER FAUCHÈRE, J. M., 2000: *Technological gap for maize (Zea mays L.) cultivation in the Polochic Watershed of Guatemala.* Ph.D. thesis. ETH Zürich, Switzerland

GARCIA, O., 1992: *Diagnostico Agronómico de los Cultivos de Maíz, Frijol y Arroz en el Valle de Polochic.* PRIAG, Informe técnico No. 3, Coronado, Costa Rica

GIGER, M., 1999: *Avoiding the Shortcut: Moving beyond the Use of Direct Incentives.* Center for Development and the Environment, Institute of Geography, University of Berne, Berne, Switzerland

GORDON, R., FRANCO, J., DE GARCIA, N., MARTINEZ, L., GONZALEZ, A., DE HERRERA, A., BOLAÑOS, J., 1993: Respuesta del maíz a la aplicación de diferentes dosis de nitrógeno en rotación con Canavalia y Mucuna bado dos tipos de labraza, Rio Hato, Panamá, 1992 - 1993, in: BOLAÑOS, J., SAIN, G., URBINA, R., BARRETO, H. (eds.). *Sintesis de Resultados Experimentales del PRM 1992*, Vol. 4, pp. 106 - 110, CIMMYT-PRM, Guatemala Ciudad, Guatemala

GRILICHES, Z., 1957: Hybrid Corn: An Exploration in the Economics of Technical Change, *Econometrica* 25: 501-22

GROHS, F., 1994: *Economics of Soil Degradation, Erosion and Conservation: A Case Study of Zimbabwe.* Ph.D. thesis, University of Hohenheim, Stuttgart, Germany

HAYAMI, Y., RUTTAN, V.,W., 1971: *Agricultural Development: An International Perspective.* 1st ed., The Johns Hopkins University Press, Baltimore, USA

HAYAMI, Y., RUTTAN, V.,W., 1985: *Agricultural Development: An International Perspective.* Rev. ed., The Johns Hopkins University Press, Baltimore, USA

HERNÁNDEZ JIMENEZ, M., SILVA GOMEZ, A. (eds.), 1994: *Diagnostico agro-socioeconomico de Polochic y Baja Verapaz, Guatemala.* PRIAG, Documento técnico No. 6, Coronado, Costa Rica

HERNANDEZ NAVAS, J., R., LAZO MELENDEZ, R., BUENO ALFERES, J., 1994: Soil Conservation Projects in El Salvador, in: LUTZ, E., PAGIOLA, S., REICHE, C. (eds.), *Economic and Institutional Analyses of Soil Conservation Projects in Central America and the Caribbean.* World Bank Environment Paper No. 8, The World Bank, Washington, D.C., USA

HICKS, J.R., 1932: *The Theory of Wages*. Macmillan & Co., London, U.K.

HONDURAS, 1993, 1994: Agricultural Census, Various Volumes, in: THORPE, A., 1997: Sustainable Agriculture in Central America, in: DEGROOT, J.P.,

RUEBEN, R. (eds.), *Sustainable Agriculture in Central America.* MacMillan Press Ltd., Houndsmills, UK

HOWARD-BORJAS, H.,1995: *Cattle and Crisis: The Genesis of Unsustainable Development in Central America.* Land Reform, FAO, Rome, Italy, pp. 87 – 116

HUSSAIN, S. S., no date: *The economics of innovation 'lock-in': An investigation into the factors that hinder the development of sustainable technological innovations in UK agriculture.* Unpublished manuscript.

IICA-Holanda/LADERAS, 1995: *Resumen Analitico de Diagnosticos Agroeconomicos de Nueva Concepción, Chalatenango.* IICA-Holanda/ LADERAS C.A., San Salvador, El Salvador

ISRIC/UNEP, 1991: *Global Assessment of the Status of Human-induced Soil Degradation*, http://www. isric.nl/GLASOD.htm

JÖRESKOG, K. G., 1993: Testing Structural Equation Models. In: BOLLEN, K. A., LONG, J. S. (eds), *Testing Structural Equation Models.* Sage Pub. Newbury Park, USA

KAIMOWITZ, D., 1996: *Livestock and Deforestation, Central America in the 1980s and 1990s: Policy Perspective.* CIFOR, Jakarta, Indonesia

KAIMOWITZ, D., 1997. Policies Affecting Deforestation for Cattle in Central America, in: DEGROOT, J.P., RUEBEN, R. (eds.), *Sustainable Agriculture in Central America.* MacMillan Press Ltd., Houndsmills, UK

KENNEDY, P. 1992: *A Guide to Econometrics.* 3rd ed, Blackwell, Oxford, UK

KREBS, M. 1999: *Erosionsanfälligkeit von ausgewählten Maisanbausystemen in Guatemala.* Semesterarbeit. Abt. VII Agrarwissenschaften, ETH Zürich, Switzerland

LAL, R., 1983: Erosion Caused Productivity Decline in Soils of Humid Tropics. *Soil Taxonomy News*, No. 5, pp. 4, 5, 18

LAL, R., 1984: Soil Erosion from Tropical Arable Lands and Its Control. *Advances in Agronomy*, No. 37, pp. 183 - 248

LAL, R., 1987: Effects of soil erosion on crop productivity. *Critical Review in Plant Science*, Vol. 5, No. 4, pp. 330 - 367

LAL, R., 1995: *Erosion Control on Sloping Land with Conservation Tillage.* RELACO III, Costa Rica

LATHWELL, D.J., 1990: Legume Green Manures: Principles for Management Based on Recent Research. *TropSoils Technical Bulletin*, No. 90 – 01, Soil Management Collaborative Research Support Program, Raleigh, NC, USA

LEONARD, H. J., 1986: *Recursos naturales y desarrollo economico en America Central, Un perfil ambiental regional.* CATIE/IIED, Turrialba, Costa Rica

LOBO BURLE, M., SUHET, A.R., PEREIRA, J., RESCK, D.V.S., PERES, J. R.R., CRAVO, M.S., BOWEN, W., BOULDIN, D.R., LATHWELL, D.J., 1992: Legume Green Manures: Dry-season Survival and the Effect on Succeeding Maize Crops. *Soil Management CRSP Bulletin* No. 92 -04, Soil Management Collaborative Research Support Program, Raleigh, NC, USA

LOEHLIN, J. C., 1998: *Latent Variable Models.* 3rd ed., Lawrence Erlbaum Associates, Mahwah, NJ, USA

LUTZ, E., PAGIOLA, S., REICHE, C., 1994: Lessons from Economic and Institutional Analyses of Soil Conservation Projects in Central America and the Caribbean, in: LUTZ, E., PAGIOLA, S., REICHE, C. (eds.), *Economic and*

Institutional Analyses of Soil Conservation Projects in Central America and the Caribbean. World Bank Environment Paper No. 8, The World Bank, Washington, D.C., USA

MADDALA, G. S., 1983: *Limited-dependent and Qualitative Variables in Econometrics.* Econometric Society Monographs, Cambridge University Press, UK

MARUYAMA, G. M., 1998: *Basics of Structural Equation Modeling.* Sage Publ., Thousand Oaks, USA

MERCADO, J., RODRÍGUEZ SANDOVAL, R., 1996: *Diagnostico Agroecológico de Nueva Concepción.* IICA-Holanda/LADERAS C.A., San Salvador, El Salvador

NAPIER, T.L., SOMMERS, D.G., 1993: Soil Conservation in the Tropics: A Prerequisite for Societal Development, in: BAUM, E., WOLFF, P, ZÖBISCH, M.A., *Acceptance of Soil and Water Conservation.* DITSL, Witzenhausen, Germany

NEILL, S. P., 1999: *Adoption and Abandonment of sustainable agriculture: The Mucuna-Maize System of Northern Honduras.* M.Sc. Thesis, Cornell University, Ithaca, USA

NUPPENAU, E.-A., HEDDEN-DUNKHORST, B., no date: *Critical Links between Food Security, Sustainable Practices and Institutions: A Causality Analysis on Smallholder Agriculture in South Africa.* Unpublished Manuscript.

OBANDO E., M., MONTALVAN, D., 1994: Technical and Economic Analysis of a Soil Conservation Project in Nicaragua, in: LUTZ, E., PAGIOLA, S., REICHE, C. (eds.), *Economic and Institutional Analysis of Soil Conservation Projects in Central America and the Caribbean.* World Bank Environment Paper No. 8, The World Bank, Washington, D.C., USA

PAGIOLA, S., 1994: Cost-Benefit Analysis of Soil Conservation, in: LUTZ, E., PAGIOLA, S., REICHE, C. (eds.), *Economic and Institutional Analyses of Soil Conservation Projects in Central America and the Caribbean.* World Bank Environment Paper No. 8, The World Bank, Washington D.C., USA

PEREIRA DE HERRERA, A., SAIN, G., 1999: *Adoption of Maize Conservation Tillage in Azuero, Panama.* CIMMYT Economics Working Paper No. 99 - 01, México, D.F., México

PINDYCK, R. S., RUBINFELD, D. L., 1981: *Econometric Models and Economic Forecasts.* 2nd ed., McGraw-Hill, London. UK

PINGALI, P., under review: Changing Locus of Agricultural Research; How will the Poor Fare?. *Food Policy*

RAVINDRAN, V., RAVINDRAN, G. 1988: Nutritional and anti-nutritional characteristics of Mucuna (*Mucuna utilis*) bean seeds. *Journal of Food Science*, 46, p. 71 - 79

RICKSON, S.T., RICKSON, R.E., NOWAK, P.J., 1993: Farmer Response to Land Degradation: Dilemmas in Sustaining Land Resources in Tropical Agriculture, in: BAUM, E., WOLFF, P., ZOBISCH, M.A. (eds.), *Acceptance of Soil and Water Conservation Strategies and Technologies*. DITSL, pp. 77 - 102, Witzenhausen, Germany

ROGERS, E., 1983: *Diffusion of innovations.* 3rd ed., Free Press, New York

ROSADO P. P., BARRIENTOS, L.E., LIMA L., S.A., 1994: Soil Erosion Control Efforts in Guatemala, in: LUTZ, E., PAGIOLA, S., REICHE, C. (eds.), *Economic and Institutional Analyses of Soil Conservation Projects in Central America and the Caribbean.* World Bank Environment Paper No. 8, The World Bank, D.C., USA

RUTHENBERG, H., 1980: *Farming Systems in the Tropics.* 3rd ed., Clarendon Press, Oxford, UK

SAIN, G., BARRETO, H., 1996: The adoption of soil conservation technology in El Salvador: Linking productivity and conservation. *Journal of Soil and Water Conservation*, Vol. 51, No. 4, pp. 313 - 321

SAIN, G., BUCKLES, D., 1998: *An Economic Analysis of the Abonera Maize Production System in the Atlantic Coast of Honduras.* CIMMYT Economics Working Paper 98 - 02. CIMMYT, El Batán, Mexico

SCHEFFER, SCHACHTSCHNABEL, P. (eds.), 1998: *Lehrbuch der Bodenkunde.* 14th ed., Ferdinand Enke Verlag, Stuttgart, Germany

SCHOOKLER, J., 1966: *Invention and Economic Growth.* Harvard University Press, Cambridge, USA

SCHRADER, K., no date. *El papel de los incentivos en la adopción de tecnologías de agricultura sostenible. El caso Guajiquiro y Opatoro, La Paz, Honduras.* PASOLAC, Managua, Nicaragua

SCHULTZ, T.P., 1964: *Transforming Traditional Agriculture.* Yale University Press, New Haven, USA

SCHUMACKER, R. E., LOMAX, R. G., 1996: *A Beginner's Guide to Structural Equation Modeling.* Lawrence Erlbaum Assoc., Mahwah, NJ, USA

STATISTICA Electronic Manual, Version 6.0, StatSoft, accessed May 22/2002 http://www.statsoft.com/textbook/stathome.html

TEMPLETON, S.R., SCHERR, S.J., 1999: Effects of Demographic and Related Microeconomic Change on Land Quality in Hills and Mountains of Developing Countries. *World Development*, Vol. 26, No. 6, pp. 903 – 918

THORPE, A., 1997: Macroeconomic Conditions for Sustainable Agriculture in Central America, in: DEGROOT, J.P., RUEBEN, R. (eds.), *Sustainable Agriculture in Central America.* MacMillan Press Ltd., Houndsmills, UK

TRIOMPHE, B. L., 1996: *Seasonal Nitrogen Dynamics and Long-term Changes in Soil Properties under the Mucuna/Maize Cropping System on the Hillsides of Northern Honduras.* Ph.D. thesis, Cornell University, Ithaca, USA

UNIVERSITY of TEXAS. Guatemala, shaded relief. Perry-Castañeda Library, Map Collection, www.lib.utexas.edu/maps/guatemala.html

UNIVERSITY of TEXAS. El Salvador, shaded relief. Perry-Castañeda Library, Map Collection, www.lib.utexas.edu/maps/elsalvador.html

UTTING, P. 1996: *Bosques, sociedad y poder.* UCA, Managua, Nicaragua

UTTING, P. 1997: Deforestation in Central America: Historical and Contemporary Dynamics, in: DEGROOT, J.P., RUEBEN, R., (eds.) *Sustainable Agriculture in Central America.* MacMillan Press Ltd., Houndsmills, UK

VAN NOORDWIJK, M., SITOMPUL, S.M., HAIRIAH, K., LISTYARINI, E., MS, S., 1995: Nitrogen supply from rotational or spatially zoned inclusion of Leguminosae for sustainable maize production on a acid soil in Indonesia, in: DATE, R.A. (ed.), *Plant Soil Interaction at low pH.* Kluwer Academic Press, Netherlands

VANCLAY, F., LAWRENCE, G., 1994: Farmer Rationality and the Adoption of Environmentally Sound Practices; A Critique of the Assumptions of Traditional Agricultural Extension. *Journal of Agricultural Education and Extension,* Vol. 1, No. 1

VAZQUEZ, A.V., 1986: La erosión y conservación del suelo en México: Realidades y Perspectivas. *Terra* 4, pp. 158 - 179

VOSTI, S. A., REARDON, T., 1997: Introduction: The critical links among sustainability, growth and poverty alleviation, in: VOSTI, S. A., REARDON, T. (eds.), *Sustainability, growth and poverty Alleviation.* The Johns Hopkins University Press, Baltimore, USA

WIGGINS, S., 1981: A Case Study in Cost-Benefit Analysis: Soil Conservation in the in the Rio Acelhuate River Basin, El Salvdor, in: MORGAN, R. (ed.) *Soil Conservation.* John Wiley and Sons Publ., Chichester, UK

WILD, A., 1995: *Umweltorientierte Bodenkunde: Eine Einführung.* Spektrum Akademischer Vlg., Heidelberg, Germany

WISCHMEIER, W. H., SMITH, D.D., 1978: Predicting Rainfall Erosion Losses: A Guide to Conservation Planning. *Agricultural Handbook* 537, U.S. Department of Agriculture, Washington, D.C., USA

WOOD, S., SEBASTIAN, K., SCHERR, S.J., 2000: *Agro-ecosystems.* IFPRI/WRI, Washington, D.C., USA

WORLD COMMISSION ON THE ENVIRONMENT AND DEVELOPMENT, 1987: *Our Common Future – The Brundtland Report.* Oxford University Press, Oxford, U.K.

ZEA, J.L., BARRETO, H., SAIN, G., BOLAÑOS, J., RAUN, W.R., 1991: Efecto de intercalar leguminosas a diferented dosis de fosforo sobre el rendimiento de maíz (*Zea mays* L.) en 24 ensayos a traves de Centro America, in: BOLAÑOS, J. (ed.), *Analisis de los Ensayos Regionales de Agronomía.* CIMMYT-PRM, Guatemala Ciudad, Guatemala

ZUREK, M., KREBS, M., 1999: La degradación del suelo en sistemas de producción de maíz en el Valle de Polochic, Guatemala. Presentation at the CIMMYT/ICTA/PRM Farmers Workshop, May 1999, Panzós, Guatemala

ZUREK, M., SAIN, G., forthcoming. *Tecnologías para la conservación del suelo en Central América. Respuestas a la demanda o a la oferta?.* CIMMYT Economics Program Working Paper, El Batan, Mexico

Annex 1

The questionnaire shown here is for surveying farmers' behavior in the second growing season (*Segunda*) in 1997 in the Polochic Valley, Guatemala. For surveying the first growing season (*Primera*) of 1997 farmers were asked in section 4 about their cropping system in this season. All the questions in the other sections were not changed.

QUESTIONNAIRE ABOUT THE ADOPTION OF SOIL CONSERVATION TECHNOLOGY AND THE MAIZE PRODUCTION SYSTEMS IN THE POLOCHIC VALLEY, DEPT. OF ALTA VERAPAZ, GUATEMALA
MARCH/APRIL 1998
(Segunda)

Name of Interviewer: ______________________________
Date: ___________ Number of questionnaire: _______

Name of Farmer: ____ (I)
Name of Wife: ____(II)
Name of Community: __________________________________(III)
Location of Community: Hillside Plains Floodplains(IV)
Distance to La Tinta/Telemán/Panzós: *(Km or time by foot)* _____________(V)

Section 1: Farmers Family

1.

	Age	Legal status*	Education (*years*)		How many month did you live on your farm in 1997?
			Primary School	Secondary School	
Farmer	(VI)	(VII)	(VIII)	(IX)	(X)

**Legal status: 1 = single; 2 = married; 3 = divorced; 4 = widowed, 5 = other (specify)*

Family members	Sex		Age		How many months did you live on the farm in 1997?	Did you help with the field work in 1997?	
	M	F	<15	>15		Yes	No
Wife		(XI)		(XII)	(XIII)		(XIV)
Child		(XV)		(XVI)	(XVII)		(XVIII)
Child		(XIX)		(XX)	(XXI)		(XXII)
Child		(XXIII)		(XXIV)	(XXV)		(XXVI)
Child		(XXVII)		(XXVIII)	(XXIX)		(XXX)
Child		(XXXI)		(XXXII)	(XXXIII)		(XXXIV)
Child		(XXXV)		(XXXVI)	(XXXVII)		(XXXVIII)
Child		(XXXIX)		(XL)	(XLI)		(XLII)
Child		(XLIII)		(XLIV)	(XLV)		(XLVI)
Other:		(XLVII)		(XLVIII)	(XLIX)		(L)
Other:		(LI)		(LII)	(LIII)		(LIV)

Section 2: Farm Resources

2. How many fields did you have in 1997? ________ (LV)

Field	Crop in *primera* 1997**** *(May/June)*	Crop in *segunda* 1997**** *(Oct./Nov.)*	Crop in the summer 1997**** *(Jan)*	Field Size *(mz - tareas)*	Location*	Inclination *(classification)*	Soil Color	Soil Quality***	Tenancy *****
1	(LVI)	(LVII)	(LVIII)	(LIX)	(LX)	(LXI)	(LXII)	(LXIII)	(LXIV)
2	(LXV)	(LXVI)	(LXVII)	(LXVIII)	(LXIX)	(LXX)	(LXXI)	(LXXII)	(LXXIII)
3	(LXXIV)	(LXXV)	(LXXVI)	(LXXVII)	(LXXVIII)	(LXXIX)	(LXXX)	(LXXXI)	(LXXXII)
4	(LXXXIII)	(LXXXIV)	(LXXXV)	(LXXXVI)	(LXXXVII)	(LXXXVIII)	(LXXXIX)	(XC)	(XCI)
5	(XCII)	(XCIII)	(XCIV)	(XCV)	(XCVI)	(XCVII)	(XCVIII)	(XCIX)	(C)

**Location: 1 = hillside, in higher part of the hill; 2 = hillside, in lower part of the hill; 3 = plains; 4 = flood plains*
** *Soil color: 1 = black, 2 = red, 3 = yellow, 4 = other (specify)*
*** Soil quality: 1 = good, 2 = regular, 3 = bad, 4 = other (specify)
**** *Crops: 1 = Maize, 2 = Beans, 3 = Rice, 4 = Velvetbean, 5 = Coffee, 6 = Cardamom, 7 = Achiote, 8 = Sorghum, 9 = other (specify)*
***** *Tenancy: 1 = registered at the* Registro General de Propiedad, *2 = INTA-Tutelaje, 3 = INTA-Titulsción Supeltcria, 4 = with a bill, 5 = indigenous community or* Parcelidad *with temporal right of use, 6 = indigenous community or Parcelidad with unlimited right of use, 7 = indigenous community or Parcelidad with heritage rights, 8 = indigenous community or Parcelidad with right to sell,9 = without registration in the* Registro General, *10 = other (specify)*

3. How many fields did you have or did you use as fallow, forest, pasture, etc. in 1997? ________ (CI)

Field	Fallow	Forest	Natural Pasture	Sown Pasture	Other (specify)
1	(CII)	(CIII)	(CIV)	(CV)	(CVI)
2	(CVII)	(CVIII)	(CIX)	(CX)	(CXI)
3	(CXII)	(CXIII)	(CXIV)	(CXV)	(CXVI)
4	(CXVII)	(CXVIII)	(CXIX)	(CXX)	(CXXI)
5	(CXXII)	(CXXIII)	(CXXIV)	(CXXV)	(CXXVI)

4. Did you have any animals in 1997 and 1996? *(Yes - No)* ___________ CXXVII)
(If "No" question 6)

5. ¿**If "Yes"**, which one did you have.? ¿How many?

Type of Animal	Number in 1997	Number of sales in 1997	Did you use maize as animal fodder? How much did you feed all of them? *(quantity)*	Number in 1996	Sale of Milk *(l)*
(CXXVIII)	(CXXIX)	(CXXX)		(CXXXI)	(CXXXII)
(CXXXIII)	(CXXXIV)	(CXXXV)		(CXXXVI)	(CXXXVII)
(CXXXVIII)	(CXXXIX)	(CXL)		(CXLI)	(CXLII)
(CXLIII)	(CXLIV)	(CXLV)		(CXLVI)	(CXLVII)
(CXLVIII)	(CXLIX)	(CL)		(CLI)	(CLII)
(CLIII)	(CLIV)	(CLV)		(CLVI)	(CLVII)

**Type of Animal: 1 = Chicken, 2 = Pigs, 3 = Cows for milk, 4 = Cows for meat, 5 = other (specify)*

6. Which machinery do you have? How many?

Type of machinery	Number
(CLVIII)	(CLIX)
(CLX)	(CLXI)
(CLXII)	(CLXIII)
(CLXIV)	(CLXV)

7. Did you or other family members who live on the farm work outside of the farm in 1997? *(Yes - No)* _______(CLXVI)
(If "No" continue with section 3)

8. **If "Yes"**, where did you work?

Family member	Type of work	How many weeks did you work outside of the farm?
1	(CLXVII)	(CLXVIII)
2	(CLXIX)	(CLXX)
3	(CLXXI)	(CLXXII)
4	(CLXXIII)	(CLXXIV)
5	(CLXXV)	(CLXXVI)
6	(CLXXVII)	(CLXXVIII)
7	(CLXXIX)	(CLXXX)

Section 3: Soil Conservation Techniques

I. Erosion

9. Which are the three biggest problems in your crops?

1. ____(CLXXXI)
2. ____(CLXXXII)
3. ____(CLXXXIII)

If the framer does not mention soil erosion….

10. In general, do you have soil erosion problems? *(Yes - No)* ________(CLXXXIV)
(Si "No" pregunta 12)

11. **If "Yes",** which are the effects of the soil erosion? _______________(CLXXXV)
(1= more stones, 2 = more landslides, 3=smaller yields, 4 = el suelo está más seco, 5 = otros (especifique))

II. The Velvetbean

12. Did you sow velvetbean in the last ten years? **(Yes - No)**________*(CLXXXVI)*
(If "No" question 34)

13. **If "Yes",** in which field(s) did you sow velvetbean in 1997? *(Number of fields of table 2)* ____________________(CLXXXVII)

If the farmer did NOT sow velvetbeans in 1997 continue with question 35.
If he used vevetbean in all maize fields continue with question 16.

14. If the farmer did not sow velvetbean in all maize fields in 1997, ….Why didn't you sow velvetbean in the other fields(s)?

____(CLXXXVIII)
(1 = could not get the seeds, 2 = there are no soil fertility problems, 3 = there are no erosion problems, 4 = does not grow in other fields, 5 = other (specify)

15. When did you sow velvetbean the last time in the other field(s)?

Field	Date
(CLXXXIX)	(CXC)
(CXCI)	(CXCII)
(CXCIII)	(CXCIV)

Select the biggest maize field of the fields where the farmer sowed velvetbean in 1997.

16. Why did you sow velvetbean in this field? ____________________(cxcv)
(1 = fertilizer 2 = less weeds, 3 = less soil loss, 4 = higher yields, 5 = other (specify)

17. Since how many years do you sow velvetbean in this field? _________(cxcvi)

18. Do you sow velvetbean every year in this field? *(Yes – No)* ___(cxcvii) *(If **"Yes"** question 20)*

19. **If "No",** why don't sow velvetbean every year?
____(cxcviii)
____

20. Which velvetbean variety did you use in this field? *(1 = white, 2 = black)* _______(cxcix)

21. Why did you use this variety? ____________________________________(cc)
(1 = don't know another one, 2 = could not find other seeds, 3 = grows quicker, 4 = has more biomass, 5 = is easier to cultivate, 6 = other (specify)

22. Where did you find the seeds? ____________________________________(cci)
(1 = other farmer, 2 = DIGESA, ICTA, 3 = CARE, 4 = farmer guards them, 5 = other (specify)

23. How much did the seeds cost? *(Q/lb)* ______________________________(ccii)

24. How many pounds did you need for this field? __________________(cciii)

25. How many days after the maize did you sow the velvetbean? _________(cciv)

26. How many persondays did you need for sowing this field?
1. Family ______________(ccv)
2. Hired laborers ______________(ccvi)

27. How many days after the maize harvest do you cut down the velvetbean? ________(ccvii)

28. How many persondays did you need for cutting this field?

1. Family ____________(CCVIII)
2.Hired Laborers _______(CCIX)

29. Where did you obtain information about the velvetbean? Who did show you how to cultivate it?___________________________ _______________(CCX)
(1 = other farmer, 2 = DIGESA, ICTA, 3 = CARE, 4 = other (specify)

30. Which other activities are necessary for the cultivation of velvetbean?
____(CCXI)

31. What has changed in this field since you cultivate velvetbean there?
____(CCXII)
(1 = higher yields, 2 =less weeds, 3 = less soil loss, 4 = more pests, 5 = more landslides, 6 = other (specify)

32. So you know since when the people use velvetbean in this valley?
________________(CCXIII)

33. Who did introduce it? ____________________________________(CCXIV)

34. **If "No"**, why didn't you sow velvetbean? ____________________(CCXV)
(1 = more pests, 2 = more work, 3 = more landslides, 4 = does not know velvetbean, 5 = could no find seeds, 6 = other (specif) (continue with question 38)

35. **If the farmer used velvetbean before but doesn't use any more**....Why don't you use it anymore?
____(CCXVI)
(1 = more pests, 2 = too much work, 3 = more landslides, 4 = could not find seeds, 5 = need the terrain to cultivate another crop (which crop?), 6 = other (specify)

36. If the farmer needs the terrain for another crop, which one? ________(CCXVII)
(1 = maize, 2 = beans, 3 = rice, 4 = coffee, 5 = cardamon, 6 = achiote, 7 = sorghum, 8 =pasture, 9 = other (specify)

37. Why?
____(CCXVIII)
____

III. Other soil conservation techniques

38. Do you know any methods to improve the soil in case that there is any soil erosion?
(Yes - No) _____________(CCXIX) ***(If "No"** question 49)*

39. Which soil conservation technique do you know? ________________(CCXX)
(1 =live barriers with Gliricidia sepium, 2 = live barriers with grass, 3 = siembra en contornos, 4 = agroforestry systems, 5 = other (specify)

40. Did you use any of these techniques in 1997? *(Yes – No)* ____________ (CCXXI)
*(**If "No"** question 51)*

41. **If "Yes"**, in which field(s) did you use it/them? Since when?

Field	Technique	Date of Introduction
(CCXXII)	(CCXXIII)	(CCXXIV)
(CCXXV)	(CCXXVI)	(CCXXVII)
(CCXXVIII)	(CCXXIX)	(CCXXX)
(CCXXXI)	(CCXXXII)	(CCXXXIII)

If the farmer used the technique(s) in all his maize fields select the biggest field and continue with question 43.
If the farmer did use it/them in all his fields continue with question 42.

42. Why didn't you use the technique(s) in the other field(s) as well? ____(CCXXXIV)
(1 = there are no soil fertility , 2 = there are no problems with soil loss, 3 = other (specify)

43. What did you have to do to implement and maintain the technique(s)? Which materials did you need?
Technique: ________________________________(CCXXXV)

Activities	Persondays		Materials	
	Family	Hired Labor	Type	Price*(Q)*
(CCXXXVI)	(CCXXXVII)	(CCXXXVIII)	(CCXXXIX)	(CCXL)
(CCXLI)	(CCXLII)	(CCXLIII)	(CCXLIV)	(CCXLV)
(CCXLVI)	(CCXLVII)	(CCXLVIII)	(CCXLIX)	(CCL)
(CCLI)	(CCLII)	(CCLIII)	(CCLIV)	(CCLV)

Technique: ______________________________ (CCLVI)

Activities	Persondays		Materials	
	Family	Hired Labor	Type	Price*(Q)*
(CCLVII)	(CCLVIII)	(CCLIX)	(CCLX)	(CCLXI)
(CCLXII)	(CCLXIII)	(CCLXIV)	(CCLXV)	(CCLXVI)
(CCLXVII)	(CCLXVIII)	(CCLXIX)	(CCLXX)	(CCLXXI)
(CCLXXII)	(CCLXXIII)	(CCLXXIV)	(CCLXXV)	(CCLXXVI)

Techniques: ______________________________ (CCLXXVII)

Activities	Persondays		Materials	
	Family	Hired Labors	Type	Prices *(Q)*
(CCLXXVIII)	(CCLXXIX)	(CCLXXX)	(CCLXXXI)	(CCLXXXII)
(CCLXXXIII)	(CCLXXXIV)	(CCLXXXV)	(CCLXXXVI)	(CCLXXXVII)
(CCLXXXVIII)	(CCLXXXIX)	(CCXC)	(CCXCI)	(CCXCII)
(CCXCIII)	(CCXCIV)	(CCXCV)	(CCXCVI)	(CCXCVII)

44. If you would need credit to implement some of these techniques where cold you get it?

____ (CCXCVIII)

(1 = Banco del Desarrollo Rural, 2 = ACT, 3 = Savings Group of the Community, 4 = Private Moneylenders, 5 = other (specify)

45. Did you notice any changes in the field since you use the soil conservation techniques? *(Yes – No)* __________ (CCXCIX) ***(If "No" question 47)***

46. **If "Yes",** which changes did you notice? ____________________________ (CCC)

(1= higher yields, 2 = less soil loss, 3 = more soil humidity, 4 = other (specify)

47. Was there a change in the yield? *(Yes – No)* __________ (CCCI) ***(If "No" question 49)***

48. **If "Yes",** how many qq/lb more did you harvest from the main field in 1997? __________________ (CCCII)

49. What would you do to obtain more information about soil conservation techniques? *(Mark the alternatives.)*

1	Talk with the neighbor	(CCCIII)
2	Go on a field trip	(CCCIV)
3	Go to a community meeting	(CCCV)

4	Take a course	(cccvi)
5	Go to Panzós to talk with the people of the DIGESA	(cccvii)
6	Pay someone to explain	(cccviii)
7	Go outside of the Valley	(cccix)
8	Other:	(cccx)

50. Which is the most important alternative? *(Ranking)* ______________(cccxi)

(Continue with section 4)

51. **If "No"**, why didn't you use any soil conservation techniques? ______(cccxii)
(1 = there are no problems with soil loss, 2 = doesn't have information, 3 = too much work, 4 = too expensive, 5 = other (specify))

52. What would you do to obtain more information about soil conservation techniques? *(Mark the alternatives.)*

1	Talk with the neighbor	(cccxiii)
2	Go on a field trip	(cccxiv)
3	Go to a community meeting	(cccxv)
4	Take a course	(cccxvi)
5	Go to Panzós to talk with the people of the DIGESA	(cccxvii)
6	Pay someone to explain	(cccxviii)
7	Go outside of the Valley	(cccxix)
8	Other:	(cccxx)

53. Which is the most important alternative? *(Ranking)* ______________(cccxxi)

Section 4: Maize production system in the Segunda *1997 or in the Summer 1997 (sowing in Oct./Nov. 1997 or Jan. 1997)*

Here select the biggest field where the farmer cultivated maize in the Segunda *1997 or in January 1997.*
Number of the field in table 2 ________(cccxxii)

104. How much time do you need to get to the field? *(time)* __________(cccxxiii)

I. History of the field

105. When did you start to cultivate this field? ____________ (CCCXXIV)

106. What was there before? ____________________ *(1 = forest, 2 = pasture, 3 = other (specify)*

107. When was the field the last time in fallow? __________________(CCCXXV)

108. For how many years did the field stay in fallow? ___________(CCCXXVI)

109. What did you cultivate in the last years?

Year	Primera	Segunda	Summer
1996	(CCCXXVII)	(CCCXXVIII)	(CCCXXIX)
1995	(CCCXXX)	(CCCXXXI)	(CCCXXXII)
1994	(CCCXXXIII)	(CCCXXXIV)	(CCCXXXV)

Crops: 1 = maize, 2 = beans, 3 = rice, 4 = velvetbean, 5 = coffee, 6 = cardamon, 7 = achiote, 8 = Sorghum, 9 = other (specify)

110. Is this field fertil, regular or poor? _________(CCCXXXVI)*(1 = fertil, 2 = regular, 3 = poor)*

111. Is this field poorer, richer or the same than when you started cultivating it? _______________(CCCXXXVII)

(1 = poorer, 2 = richer, 3 = the same) ***(If "the same"*** *question 114)*

112. If it changed, why? ___________________________________(CCCXXXVIII)

(1 = the soil is tired 2 = more weeds, 3 = more pests, 4 = velvetbean, 5 = fertilizer, 6 = other cultivation technique, 7 = other (specify)

113. Which is the most important reason? *(Number of the question before)* ______(CCCXXXIX)

114. Doyou have problems with soil erosion in this field? *(Yes – No)* _____________(CCCXL)

115. Did you once sow velvetbean in this field? *(Yes – No)* _________(CCCXLI)

(If "No" *question 117)*

116. **If "Yes",** when was the first time? *(Date)* ___________________(CCCXLII)

II. Soil Preparation

117. What did you do to prepare the soil?

Activities	Days after planting	Number of persondays	
		Family	Hired Labor
Only cutting *(→ Question 120)*	(CCCXLIII)	(CCCXLIV)	(CCCXLV)
Burning with Herbicices *(→ Question 118)*	(CCCXLVI)	(CCCXLVII)	(CCCXLVIII)

118. **If the farmer burned with herbicides,** why did you apply herbicides?
____________(CCCXLIX)

119. Which products did you use? In which quantity?

Product	Days before/ after Planting*	Copas Bayer/ Bomb	Bombs/ Field	Number of Persondays	
				Family	Hired labor
(CCCL)	(CCCLI)	(CCCLII)	(CCCLIII)	(CCCLIV)	(CCCLV)
(CCCLVI)	(CCCLVII)	(CCCLVIII)	(CCCLIX)	(CCCLX)	(CCCLXI)
(CCCLXII)	(CCCLXIII)	(CCCLXIV)	(CCCLXV)	(CCCLXVI)	(CCCLXVII)

(Copa Bayer: 25 cm^3)
** Days before/after planting: # = number of days before planting , # + = after planting, # 0 = day of planting*

III. Planting

120. Which maize variety did you use in this field in the *segunda*1997?

	Variety 1	Variety 2
Name*	(CCCLXVIII)	(CCCLXIX)
Since how many years did you use this variety?	(CCCLXX)	(CCCLXXI)
Quantity used in this field *(specify)*	(CCCLXXII)	(CCCLXXIII)
Origin of the seeds**	(CCCLXXIV)	(CCCLXXV)
Price of the seeds	(CCCLXXVI)	(CCCLXXVII)
How much of the 1996 production of this field did you have to sell to buy the seeds for this field in 1997?	(CCCLXXVIII)	(CCCLXXIX)
If this was the farmers own seed.... when did you obtain the original seeds?	(CCCLXXX)	(CCCLXXXI)

1 = criollo, 2 = HB 83, 3 = other (specify)
***1 = own, 2 = bought at the ICTA, 3 = bought from another farmer, 4 = bought from commercial agent, 5 = other (specify)*

121. Which ist the most important advantage of this variety __________(CCCLXXXII)

____

122. Which is the most important problem of this variety? __________(CCCLXXXIII)

____

123. When did you plant? *(Date)* ________________________(CCCLXXXIV)

124. ¿How did you plant? ____________________(CCCLXXXV)*(1 = planting stick, 2 = seeder, 3 = other (specify)*

125. How many persondays did you need for planting this field?

1. Family ________(CCCLXXXVI)
2. Contratados _______(CCCLXXXVII)

126. Distance between the rows (cm) ____________________(CCCLXXXVIII)

127. Distance between plants (cm) ____________________(CCCLXXXIX)

128. Number of seeds per planting hole __________________(CCCXC)

IV. Weeds

129. How did you control the weeds in this field in the *segunda* 1997?

Weeding	Method*	Days after Planting	Persondays		Copas Bayer/ Bomb	Bomb/ Field
			Family	Hired Labor		
1	(CCCXCI)	(CCCXCII)	(CCCXCIII)	(CCCXCIV)	(CCCXCV)	(CCCXCVI)
2	(CCCXCVII)	(CCCXCVIII)	(CCCXCIX)	(CD)	(CDI)	(CDII)
3	(CDIII)	(CDIV)	(CDV)	(CDVI)	(CDVII)	(CDVIII)

**Method: 1 = manual, 2 = herbicides (which ones?), 3 = other (specify)*

V. Fertilization

130.Did you use fertilizer in this field in the segunda 1997 *(Yes - No)* ____(CDIX)
***(If "No"** question 133)*

131. Which type of fertilizer did you use?

Type of Fertilizer*	Days after planting	Quantity/ Field	Method of application**	Persondays	
				Family	Hired Labor
(CDX)	(CDXI)	(CDXII)	(CDXIII)	(CDXIV)	(CDXV)
(CDXVI)	(CDXVII)	(CDXVIII)	(CDXIX)	(CDXX)	(CDXXI)
(CDXXII)	(CDXXIII)	(CDXXIV)	(CDXXV)	(CDXXVI)	(CDXXVII)
(CDXXVIII)	(CDXXIX)	(CDXXX)	(CDXXXI)	(CDXXXII)	(CDXXXIII)

** 1 = Urea, 2 = 15-15-15, 3 = compost, 4 = other (specify)*

*** 1 = manual, in planting hole, 2 = manual, between rows, 3 = with machinery, 4= other (specify)*

132. Where did you buy the fertilizer? ____________________________(CDXXXIV)
(1 = La Tinta, 2 = Panzós, 3 = Telemán, 4 = other (specify)

VI. Pest Control

133. Did you pest control measures in this field in the *segunda* 1997? *(Yes - No)*
__________(CDXXXV)
***(If "No"** question 135)*

134. How did you control pests in your field?

Type of Insecticide	Days after Planting	Quantity/ Field	Persondays	
			Family	Hired Labor
(CDXXXVI)	(CDXXXVII)	(CDXXXVIII)	(CDXXXIX)	(CDXL)
(CDXLI)	(CDXLII)	(CDXLIII)	(CDXLIV)	(CDXLV)
(CDXLVI)	(CDXLVII)	(CDXLVIII)	(CDXLIX)	(CDL)
(CDLI)	(CDLII)	(CDLIII)	(CDLIV)	(CDLV)

VII. Harvest

135. How much did you harvest of this field in the *segunda* 1997? *(quantity)*
__________(CDLVI)

136. How many qq/lb of this harvest did you guard for the house and the animals? *(qq/lb)* __________(CDLVII)

137. How long does this last for the house? *(months)* __________________(CDLVIII)

138. How much did you give to the animals? *(qq/lb)* __________________(CDLIX)

139. If you sold a part of the harvest, where did you sell it? _________(CDLX)
(1 = on the farm, 2 = in the village, 3 = other (specify)

140. Did you harvest fresh maize as well? *(Yes - No)* ________(CDLXI) ***(If "No" question 143)***

141. **If "Yes",** how much did you harvest? *(quantity)* __________________(CDLXII)

142. How much did you use for the family? _________________________(CDLXIII)

143. In the other fields where you cultivated maize in the *segunda* 1997 did you use another variety than in the field about which we talked before? Which one did you use? __________________(CDLXIV)
(1 = HB 83, 2 = Criollo, 3 = other (specify) ***(If "No"*** *question 145)*

144. Why? ____(CDLXV)
____

145. Did you burn the fields in which you cultivated maize in the ***primera 1997***?
(Yes – No) ________(CDLXVI) ***(If "No"*** *question 147)*

146. **If "Yes",** why did you do it?
____(CDLXVII)
(1 = because you do it like this, 2 = less weeds, 3 = less rats, 4 = easier to plant, 5 = other (specify)
(continue with question 152)

147. **If the farmer did not burn his field(s),** since how many years are you not burning any more? ____________(CDLXVIII)

148. What has changed since you don't burn anymore? _______________(CDLXIX)
(1 = less oil loss, 2 = more weeds, 3 = more work for cleaning the field, 4 = more pests, 5 = higher yields, 6 = other (specify)

149. From where did you get information about "not burning"? _________ (CDLXX)
(1 = DIGESA, ICTA, 2 = CARE, 3 = other farmer, 4 = family member, 5 = other (specify)

150. What are the two most important advantages of "not burning"?
1. ______________________________ (CDLXXI)
2. ______________________________ (CDLXXII)

151. What are the two most important disadvantages of "not burning"?
1. ______________________________ (CDLXXIII)
2. ______________________________ (CDLXXIV)

152. Did you need credit to buy the inputs for this field? *(Yes - No)* ____ (CDLXXV)
***(If "No"** continue with section 5)*

153. ***If "Yes"***, which inputs did you buy with credit? ______________ (CDLXXVI)

Section 5: Credit Market

154. Who guards the money in your family? ________________ (CDLXXVII)
(1 = farmer, 2 = wife, 3 = both, 4 = children, 5 = other (specify)

155. Where can you go to obtain credit? ________________ (CDLXXVIII)
(1 = Banco del Desarrollo Rural, 2 = ACT, 3 = Savings Group of the Community, 4 = Private Moneylenders, 5 = other (specify)

156. Which are the conditions to obtain credit?

Organization	Form of credit*	Land title needed?	Guarantor needed?	Animals needed?	Annual interest rate *(%)*
Bank	(CDLXXIX)	(CDLXXX)	(CDLXXXI)	(CDLXXXII)	(CDLXXXIII)
ACT	(CDLXXXIV)	(CDLXXXV)	(CDLXXXVI)	(CDLXXXVII)	(CDLXXXVIII)
Savings Group	(CDLXXXIX)	(CDXC)	(CDXCI)	(CDXCII)	(CDXCIII)
Private Moneylender	(CDXCIV)	(CDXCV)	(CDXCVI)	(CDXCVII)	(CDXCVIII)
Other:	(CDXCIX)	(D)	(DI)	(DII)	(DIII)

**1 =Money, 2 = Inputs*

157. Are there any other conditions to obtain credit? ____________________

____

158. Did you oncc apply for credit? *(Yes – No)* ____________(DIV) *(If "No" question 162)*

159. **If "Yes"**, for what? ________________________________(DV)
(1 = fertilizer, 2 = hybrid seeds, 3 = other kind of seeds (specify), 4 = other inputs (specify), 5 = for something of the house, 6 = to hire laborers, 7 = other (specify)

160. Did you obtain the credit/ *(Yes – No)* ____________________(DVI) ***(If "Yes" section 6)***

161. **If "No"**, why not? ________________________________(DVII)
(1 = no land title, 2 = no guarantor, 3 = interest rate too high, 4 = other (specify)

162. **If "No"**, why? ________________________________(DVIII)
(1 = no landtitle, 2 = no guarantor, 3 = interest rate too high, 4 = other (specify)

Section 6: Other questions

163. Who participates in the farm activities?

Activities	Farmer	Wife	Children	Hired Labor	Other
Cleaning of field(s)	(DIX)	(DX)	(DXI)	(DXII)	(DXIII)
Planting of maize	(DXIV)	(DXV)	(DXVI)	(DXVII)	(DXVIII)
Fertilizing	(DXIX)	(DXX)	(DXXI)	(DXXII)	(DXXIII)
Weeding	(DXXIV)	(DXXV)	(DXXVI)	(DXXVII)	(DXXVIII)
Pest control	(DXXIX)	(DXXX)	(DXXXI)	(DXXXII)	(DXXXIII)
Maize harvest	(DXXXIV)	(DXXXV)	(DXXXVI)	(DXXXVII)	(DXXXVIII)
Taking the grains off	(DXXXIX)	(DXL)	(DXLI)	(DXLII)	(DXLIII)
Coffee harvest	(DXLIV)	(DXLV)	(DXLVI)	(DXLVII)	(DXLVIII)
Animals	(DXLIX)	(DL)	(DLI)	(DLII)	(DLIII)
Housework	(DLIV)	(DLV)	(DLVI)	(DLVII)	(DLVIII)
Care of Children	(DLIX)	(DLX)	(DLXI)	(DLXII)	(DLXIII)
Get firewood	(DLXIV)	(DLXV)	(DLXVI)	(DLXVII)	(DLXVIII)
Communal/social/religious activities	(DLXIX)	(DLXX)	(DLXXI)	(DLXXII)	(DLXXIII)
Other:	(DLXXIV)	(DLXXV)	(DLXXVI)	(DLXXVII)	(DLXXVIII)
Other:	(DLXXIX)	(DLXXX)	(DLXXXI)	(DLXXXII)	(DLXXXIII)

164. Who decides when and how to work in the fields? ______________(DLXXXIV)
(1 = farmer, 2 = wife, 3 = both, 4 = other (specify)

165. Do you know some farmers' organizations? *(Yes – No)* _________(DLXXXV)
***(If "No"** question 171)*

166. **If " Yes",** which are they?
____(DLXXXVI)
____

167. Is there any other working group with farmer or other community members? *(Yes – No)* _________ (DLXXXVII)
***(If "No"** question 171)*

168. Do you participate in some of these groups or organizations? *(Yes – No)* _________(DLXXXVIII)
***(If "No"** question 171)*

169. **If "Yes",** in which group? ______________________________(DLXXXIX)

170. What do you do in this group? ______________________________(DXC)
____

171. Where can you get information about new farming practices to improve your fields?
____(DXCI)
____

172. Which products did you sell in the year 1997?

Product	Quantity *(qq - lb)*	Price
(DXCII)	(DXCIII)	(DXCIV)
(DXCV)	(DXCVI)	(DXCVII)
(DXCVIII)	(DXCIX)	(DC)
(DCI)	(DCII)	(DCIII)
(DCIV)	(DCV)	(DCVI)
(DCVII)	(DCVIII)	(DCIX)

173. Who decides when and how to sell? ______________________________(DCX)
(1 = the farmer, 2 = the wife, 3 = both, 4 = other (specify)

174. If you want to sell something of your farm how do you decide how much you can sell?

____(DCXI)

____

Section 7: Questions for the farmer's wife or another female household member

175. In which farm activities do you participate?

Activities	Wife
Cleaning of field(s)	(DCXII)
Planting of maize	(DCXIII)
Fertilizing	(DCXIV)
Weeding	(DCXV)
Pest control	(DCXVI)
Maize harvest	(DCXVII)
Taking the grains off	(DCXVIII)
Coffee harvest	(DCXIX)
Animals	(DCXX)
Housework	(DCXXI)
Care of Children	(DCXXII)
Get firewood	(DCXXIII)
Communal/social/religious activities	(DCXXIV)
Other:	(DCXXV)
Other:	(DCXXVI)

176. Does your husband talk with you about the farm activities? *(Yes - No)* ___________(DCXXVII)

177. Do you know if there are problems with soil loss on the farm? *(Yes - No)* ________(DCXXVIII)

178. Do you know any methods to improve the soil if there has been any soil loss?

____(DCXXIX)

179. Do you know the velvetbean? *(Yes - No)* ___________(DCXXX) ***(If "No"*** *question 191)*

180. **If "Yes"**, did your husband sowed velvetbean in 1997? (Yes - *No)* __________(DCXXXI) ***(If "No"** question 186)*

181. **If "Yes"**, do you sometimes help your husband with the work in the velvetbean? *(Yes - No)* _____________(DCXXXII) ***(If "No"** question 183)*

182. **If "Yes"**, what do you do? ________________________________(DCXXXIII)
(1 = planting, 2 = guard the seeds, 3 = buy the seeds, 4 = other (specify)

183. What changes have you noticed in the field(s) where you planted the velvetbean? (Yes - No) _______*(DCXXXIV)*
***(If "No"** question 185)*

184. **If "Yes"**, which changes did you notice? ____________________(DCXXXV)
(1 = higher yields, 2 = less weeds, 3 = less soil loss, 4 = more pests, 5 = more landslides, 6 = other (specify)

185. From where did you obtain the information about the velvetbean? Where did you learn how to cultivate it?
____(DCXXXVI)
(1 = other farmer, 2 = DIGESA, ICTA, 3 = CARE, 4 = other (specify)
(continúe con pregunta 191)

186.**If "No"**, did your husband use them before? *(Yes - No)* __________(DCXXXVII)
***(If "Yes"** question 187, **if "No"** question 190)*

187. **If the farmer used velvetbean but doesn't use it anymore** why not?
____(DCXXXVIII)
(1 = more pests, 2 = too much work, 3 = more landslides, 4 = couldn't obtain the seeds, 5 = needs the terrain to cultivate other crops (which corps?), 6 = other (specify)

188. If the farmer needs the terrain to cultivate another crop, which one do you cultivate? _________________(DCXXXIX)
(1 = maize, 2 = beans, 3 = rice, 4 = coffee, 5 = cardamon, 6 = achiote, 7 = sorghum, 8 = other (specify)

189. Why? ____(DCXL)
(continue with question 191)

190. **If "No"**, why doesn't your husband use velvetbean? _____________(DCXLI)
(1 = more pests, 2 = more work, 3 = more landslides, 4 = doesn't know velvetbean, 5 = other (specify)

191. Which soil conseravtion technology do you know? ______________(DCXLII)
= live barriers with Gliricidai sepium, 2 = live barriers with grass, 3 = siembra en contornos, 4 = agroforetry systems, 5= (specify)

192. Did your husband use any of these techniques in 1997? *(Yes – No)* ____(DCXLIII) ***(If "No"** question 197)*

193. **If "Yes"**, which ones did he use? *(Note the number(s) of question 191)*______________ (DCXLIV)

194. Do you help him with the activities of these techniques? *(Yes – No)* __________(DCXLV) ***(If "No"** question 196)*

195. **If "Yes"**, what do you do? ________________________________(DCXLVI)

____

196. From where did you obtain the information about these techniques?

No. de Technique	Source of information
(DCXLVII)	(DCXLVIII)
(DCXLIX)	(DCL)
(DCLI)	(DCLII)
(DCLIII)	(DCLIV)

**1 = other farmer, 2 = DIGESA, ICTA, 3 = CARE, 4 = other (specify)*

197. **If "No"**, why doesn't your husband use any of the soil conservation techniques? ___(DCLV

____

(1 = there are no soil loss problems, 2 = doesn't have information, 3 = too much work, 4 = to expensive, 5 = other (specify)

198. Do you any groups or organizations in the community? *(Yes – No)* ________(DCLVI) ***(If "No"** question 202)*

199. **If "Yes"**, which one so you know?

____(DCLVII)

____

200. Do you participate in any of them? *(Yes – No)* ________(DCLVIII) ***(If "No"** question 202)*

201. **If "Yes"**, what do you do in this group?

____(DCLIX)

____

202. Where can your husband obtain information about new farming practices to improve his fields?

____(DCLX)

Section 8: Observations

General: ____

____

____

____

____

____

The farmer: ____

____

____

____

____

____

____

The wife: ____

____

____

____

Annex 2

ENCUESTA

Origen y Desarrollo de las Técnicas de Conservación del Suelo en Nueva Concepción, El Salvador

(ORGANIZACIONES)

Nombre del representante entrevistado: ______________________________[DCLXI]
Posición del representante: ______________________________[DCLXII]
Nombre de la institución: ______________________________[DCLXIII]
Dirección: ______________________________[DCLXIV]
Teléfono: ______________________________[DCLXV]

1. Perfil de la Organización

1. Tipo de organización: Pública [DCLXVI]:❒, ONG [DCLXVII]: ❒, Proyecto [DCLXVIII]: ❒, Programa [DCLXIX] : ❒, Organización [DCLXX]: ❒, Otro [DCLXXI]: ❒ __________________

2. En qué cantones o caseríos de Nueva Concepción trabaja su institución, proyecto u organización?
1. [DCLXXII]
____
2. [DCLXXIII]
____
3. [DCLXXIV]
____

3. Porqué seleccionó su institución Nueva Concepción para trabajar (razones o criterios)?

1.

DCLXXV

2.

DCLXXVI

3.

DCLXXVII

4. Cuántas personas en total trabajan en su institución: ______________ DCLXXVIII

5. Cuántas personas trabajan en Nueva Concepción: _________________ DCLXXIX

6. Cuántas de estas personas se dedican al área de conservación de suelo:
______ DCLXXX

7. Desde qué año trabaja su institución en el área de conservación de suelos :
El Salvador: ___________________ DCLXXXI Nueva Concepción: _________ DCLXXXII

8. Cuáles son las actividades y áreas en que está trabajando su institución y el porcentaje de participación de cada una de ellas.

Tipo de actividades/áreas	**Porcentaje (%)**	**Areas Conservación %**
DCLXXXIII	DCLXXXIV	DCLXXXV
DCLXXXVI	DCLXXXVII	DCLXXXVIII
DCLXXXIX	DCXC	DCXCI
DCXCII	DCXCIII	DCXCIV
DCXCV	DCXCVI	DCXCVII
DCXCVIII	DCXCIX	DCC
DCCI	DCCII	DCCIII
DCCIV	DCCV	DCCVI
DCCVII	DCCVIII	DCCIX
DCCX	DCCXI	DCCXII
DCCXIII	DCCXIV	DCCXV

9. De dónde obtiene su organización información sobre las diferentes técnicas de conservación de suelos?

1. ______________________________ DCCXVI
2. ______________________________ DCCXVII
3. ______________________________ DCCXVIII

10. De que manera distribuyen ustedes esa información al resto de los miembros de su organización?

1. ______________________________ DCCXIX
2. ______________________________ DCCXX
3. ______________________________ DCCXXI

2. El Trabajo en la Municipalidad de Nueva Concepción

11. Qué otras instituciones han trabajado en obras de conservación en Nueva Concepción?

Institución	**Técnicas de Conservación**	**Años (Inicio y Final)**
DCCXXII	DCCXXIII	DCCXXIV
DCCXXV	DCCXXVI	DCCXXVII
DCCXXVIII	DCCXXIX	DCCXXX
DCCXXXI	DCCXXXII	DCCXXXIII
DCCXXXIV	DCCXXXV	DCCXXXVI
DCCXXXVII	DCCXXXVIII	DCCXXXIX
DCCXL	DCCXLI	DCCXLII

2.A Identificación de los Problemas

10. Cómo identificó su institución los problemas de suelo en Nueva Concepción (Metodología : sondeos, encuestas, diagnósticos, talleres participativos, etc.)?

Metodología Identificación de los Problemas	Instituciones Participantes (especifique)					
	Comunidades	Org. Gubernamentales	Consultores	ONG	Proyectos	Otros
DCCXLIII	DCCXLIV	DCCXLV	DCCXLVI	DCCXLVII	DCCXLVIII	DCCXLIX
DCCL	DCCLI	DCCLII	DCCLIII	DCCLIV	DCCLV	DCCLVI
DCCLVII	DCCLVIII	DCCLIX	DCCLX	DCCLXI	DCCLXII	DCCLXIII
DCCLXIV	DCCLXV	DCCLXVI	DCCLXVII	DCCLXVIII	DCCLXIX	DCCLXX
DCCLXXI	DCCLXXII	DCCLXXIII	DCCLXXIV	DCCLXXV	DCCLXXVI	DCCLXXVII
DCCLXXVIII	DCCLXXIX	DCCLXXX	DCCLXXXI	DCCLXXXII	DCCLXXXIII	DCCLXXXIV
DCCLXXXV	DCCLXXXVI	DCCLXXXVII	DCCLXXXVIII	DCCLXXXIX	DCCXC	DCCXCI

2.B Selección de las Técnicas de Conservación

11. Cuáles fueron los criterios para seleccionar las técnicas de conservación promovidas en Nueva Concepción? ¿De dónde obtuvo la información sobre estas técnicas?

Técnica seleccionada	**Criterios para la selección**	**Fuentes de información**
DCCXCII	DCCXCIII	DCCXCIV
DCCXCV	DCCXCVI	DCCXCVII
DCCXCVIII	DCCXCIX	DCCC
DCCCI	DCCCII	DCCCIII
DCCCIV	DCCCV	DCCCVI

12. Cómo actualizan ustedes la información sobre las técnicas seleccionadas?
1. ______________________________ DCCCVII
2. ______________________________ DCCCVIII
3. ______________________________ DCCCIX

13. Quiénes estuvieron involucrados en la selección final de las técnicas conque su organización trabajaría en Nueva Concepción?
1. ______________________________ DCCCX
2. ______________________________ DCCCXI
3. ______________________________ DCCCXII

2.C Investigación sobre las técnicas de conservación

14. Se hizo algún tipo de investigación sobre las técnicas de conservación antes de su promoción? *(Sí/No)*: ______________________ DCCCXIII *(Si "No": Preg. 16)*

15. **Si "Sí"**, Qué tipo de investigación se hizo?

Técnicas	Tipo de Investigación	Participantes en la Investigación	Problema Investigado	Duración
DCCCXIV	DCCCXV	DCCCXVI	DCCCXVII	DCCCXVIII
DCCCXIX	DCCCXX	DCCCXXI	DCCCXXII	DCCCXXIII
DCCCXXIV	DCCCXXV	DCCCXXVI	DCCCXXVII	DCCCXXVIII
DCCCXXIX	DCCCXXX	DCCCXXXI	DCCCXXXII	DCCCXXXIII
DCCCXXXIV	DCCCXXXV	DCCCXXXVI	DCCCXXXVII	DCCCXXXVIII
DCCCXXXIX	DCCCXL	DCCCXLI	DCCCXLII	DCCCXLIII

Código para "Tipo de investigación": 1=Investigación básica, 2=Investigación adaptativa

2.D Transferencia de las Técnicas de Conservación

16. Qué métodos de transferencia utilizó su organización para promover las técnicas en Nueva Concepción?. Por cuánto tiempo las utilizó?

TÉCNICA PROMOVIDA	MÉTODO DE TRANSFERENCIA	Duración del Uso del Método de Transferencia
DCCCXLIV	DCCCXLV	DCCCXLVI
DCCCXLVII	DCCCXLVIII	DCCCXLIX
DCCCL	DCCCLI	DCCCLII
DCCCLIII	DCCCLIV	DCCCLV
DCCCLVI	DCCCLVII	DCCCLVIII

17. Utilizó su organización algún tipo de incentivos para promover el uso de las técnicas)? (Sí/No): _______________ DCCCLIX *(Si "No": Preg. 19)*

18. **Si "Sí"**,Qué tipo de incentivos utilizó su organización? ¿Por cuánto tiempo?

En caso de que la organización no siga usando alguno de los incentivos,
Por qué se terminaron de utilizar estos incentivos?

TÉCNICA PROMOVIDA	**TIPO DE INCENTIVO**	**Duración**		**Por qué se terminó**
		Inicio	**Final**	
DCCCLX	DCCCLXI	DCCCLXII	DCCCLXIII	DCCCLXIV
DCCCLXV	DCCCLXVI	DCCCLXVII	DCCCLXVIII	DCCCLXIX
DCCCLXX	DCCCLXXI	DCCCLXXII	DCCCLXXIII	DCCCLXXIV
DCCCLXXV	DCCCLXXVI	DCCCLXXVII	DCCCLXXVIII	DCCCLXXIX

19. Con respecto a las técnicas promovidas por su organización, Qué tipo de capacitación o asistencia técnica se ha brindado a los agricultores de Nueva Concepción? ¿Por cuánto tiempo se brindó esta capacitación o asistencia técnica? *En caso de que la organización no siga brindando capacitación o asistencia técnica alguna,* ¿Por qué dejó de hacerlo?

TÉCNICA PROMOVIDA	**TIPO DE ASISTENCIA TÉCNICA O CAPACITACIÓN**	**Duración**		**Por qué se terminó**
		Inicio	**Final**	
DCCCLXXX	DCCCLXXXI	DCCCLXXXII	DCCCLXXXIII	DCCCLXXXIV
DCCCLXXXV	DCCCLXXXVI	DCCCLXXXVII	DCCCLXXXVIII	DCCCLXXXIX
DCCCXC	DCCCXCI	DCCCXCII	DCCCXCIII	DCCCXCIV
DCCCXCV	DCCCXCVI	DCCCXCVII	DCCCXCVIII	DCCCXCIX
CM	CMI	CMII	CMIII	CMIV

20. Cuáles fueron los tres problemas más importantes en todo el proceso de selección de las técnicas de conservación para Nueva Concepción?
1. ____[CMV]
2. ____[CMVI]
3. ____[CMVII]

2. E. Evaluación de las Técnicas de Conservación

21. Ha hecho su organización alguna evaluación sobre la adopción de las técnicas en Nueva Concepción? (*Sí/No*) ______________[CMVIII]
(Si "No" → Final de la Encuesta)

22. **Si "Sí"**, Cuáles fueron los resultados de esta evaluación para cada técnica?
Técnica 1: ______________________[CMIX]
1. ____[CMX]
2. ____[CMXI]
3. ____[CMXII]

Técnica 2: ______________________[CMXIII]
1. ____[CMXIV]
2. ____[CMXV]
3. ____[CMXVI]

Técnica 3: ______________________[CMXVII]
1. ____[CMXVIII]
2. ____[CMXIX]
3. ____[CMXX]

Técnica 4: ______________________[CMXXI]
1. ____[CMXXII]
2. ____[CMXXIII]
3. ____[CMXXIV]

Técnica 5: ______________________[CMXXV]
1. ____[CMXXVI]
2. ____[CMXXVII]
3. ____[CMXXVIII]

23. Han habido cambios en la promoción de las técnicas después de esta evaluación? (*Sí/No*) ______________[CMXXIX] *(Si "No" → Final de la Encuesta)*

24. Si "Sí", Qué se ha cambiado con la evaluación?

1. ____[CMXXX]
2. ____[CMXXXI]
3. ____[CMXXXII]

24. *En caso de que la organización no siga promoviendo una de las técnica,* ¿Cuáles son las razones para abandonar la promoción de esta técnica?

Técnica:________________[CMXXXIII]

1. ____[CMXXXIV]
2. ____[CMXXXV]
3. ____[CMXXXVI]

Técnica:________________[CMXXXVII]

1. ____[CMXXXVIII]
2. ____[CMXXXIX]
3. ____[CMXL]

I. Observaciones del Encuestador

____
____
____
____
____
____
____
____
____
____
____

Nombre del Encuestador:____
Fecha:____________________________
Numero de Encuesta:________________

Annex 3

ENCUESTA:

Origen y Desarrollo de las Técnicas de Conservación del Suelo en Nueva Concepción, El Salvador

(AGRICULTORES)

Nombre del Encuestador:____
Fecha:__________________________ Numero de Encuesta:__________

Nombre de Cantón:_________________[CMXLI] Caserío ______________[CMXLII]
Nombre del Agricultor:____[CMXLIII]
Edad del Agricultor:_____[CMXLIV]
Años de Primaria:_______[CMXLV]Años de Secundaria:_________[CMXLVI]
En que año comenzó a trabajar con la agricultura: _________________[CMXLVII]

1. Dibujo de las parcelas:
(Haga un dibujo de todos las parcelas del agricultor con la información de ser posible)

Si en la conversación sobre las parcelas sale que **el agricultor usa alguna técnica** de conservación, sigue con la **parte 2**. (Uso de Técnicas de Conservación).

Si el agricultor realmente **no usa ninguna técnica** de conservación, sigue con la **parte 5**. **(pagina 8**, El No-Uso de Técnicas de Conservación).

2. Uso de Técnicas de Conservación

1. Tenencia de la **Tierra Propia** CMXLVIII

2. Año en que comenzó a trabajar en esta finca: ________ CMXLIX

Técnicas de Conser-vación (Tipo)	Cultivos	Epoca de Siembra de cada cultivo	Rendimiento por Cultivo	Años del uso de la técnica	Institución que promovió
CML	CMLI	CMLII	CMLIII	CMLIV	CMLV
CMLVI	CMLVII	CMLVIII	CMLIX	CMLX	CMLXI
CMLXII	CMLXIII	CMLXIV	CMLXV	CMLXVI	CMLXVII
CMLXVIII	CMLXIX	CMLXX	CMLXXI	CMLXXII	CMLXXIII

Técnicas de Conser-vación (Tipo)	Características de la Parcela (Suelo, Pendiente, Pedregosidd, Uso del Barbecho (cuándo?))	Area de la Parcela (Mz/Tareas)	Observaciones
CMLXXIV	CMLXXV	CMLXXVI	CMLXXVII
CMLXXVIII	CMLXXIX	CMLXXX	CMLXXXI
CMLXXXII	CMLXXXIII	CMLXXXIV	CMLXXXV
CMLXXXVI	CMLXXXVII	CMLXXXVIII	CMLXXXIX

3. Tenencia de la **Tierra Alquilada**[CMXC]

4. Años que comenzó a trabajar en esta finca: ____________________ CMXCI

Técnicas de Conser-vación (Tipo)	**Cultivos**	**Epoca de Siembra de cada cultivo**	**Rendi-miento por Cul-tivo**	**Años del uso de la técnica**	**Institución que promovió**
CMXCII	CMXCIII	CMXCIV	CMXCV	CMXCVI	CMXCVII
CMXCVIII	CMXCIX	M	MI	MII	MIII
MIV	MV	MVI	MVII	MVIII	MIX
MX	MXI	MXII	MXIII	MXIV	MXV
MXVI	MXVII	MXVIII	MXIX	MXX	MXXI

Técnicas de Conser-vación (Tipo)	**Características de la Parcela (Suelo, Pendiente, Pedregosidd, Uso del Bar-becho (cuándo?))**	**Area de la Parcela (Mz/Tarea s)**	**Observaciones**
MXXII	MXXIII	MXXIV	MXXV
MXXVI	MXXVII	MXXVIII	MXXIX
MXXX	MXXXI	MXXXII	MXXXIII

5. Tiene otra parcela donde **NO** tiene Técnicas de Conservación?
Propia: ____________ MXXXIV Alquilada: ____________ MXXXV

Cultivo	**Tamaño de la Parcela (mz.)**	**Razón de No-uso**
MXXXVI	MXXXVII	MXXXVIII
MXXXIX	MXL	MXLI
MXLII	MXLIII	MXLIV
MXLV	MXLVI	MXLVII

6. Porqué hace Técnicas de Conservación?

Técnica	**Razón de hacerlas**
MXLVIII	MXLIX
ML	MLI
MLII	MLIII
MLIV	MLV

7. Como conoció las Técnicas de Conservación?

Técnica usada	**Donde conoció la Técnica por primera vez**	**Tipo de evento o experiencia**	**Institución**
MLVI	MLVII	MLVIII	MLIX
MLX	MLXI	MLXII	MLXIII
MLXIV	MLXV	MLXVI	MLXVII
MLXVIII	MLXIX	MLXX	MLXXI
MLXXII	MLXXIII	MLXXIV	MLXXV

8. Capacitaciones recibidas para aprender sobre Técnicas de Conservación.

Institución	Tipo de Capacitación	Duración días	Lugar	Año
MLXXVI	MLXXVII	MLXXVIII	MLXXIX	MLXXX
MLXXXI	MLXXXII	MLXXXIII	MLXXXIV	MLXXXV
MLXXXVI	MLXXXVII	MLXXXVIII	MLXXXIX	MXC
MXCI	MXCII	MXCIII	MXCIV	MXCV

9. Tipo de ayuda (herramientas por trabajo, alimentos, plantas, insumos, asistencia técnica, etc.) por parte de las instituciones para introducción o mantenimiento de Técnicas de Conservación.

Técnica de Conservación	Tipo de Ayuda	Institución	Año de Inicio	Duración
MXCVI	MXCVII	MXCVIII	MXCIX	MC
MCI	MCII	MCIII	MCIV	MCV
MCVI	MCVII	MCVIII	MCIX	MCX
MCXI	MCXII	MCXIII	MCXIV	MCXV
MCXVI	MCXVII	MCXVIII	MCXIX	MCXX

3. Manejo de las Técnicas de Conservación de Suelos

10. Se ha cambiado la manera en que usted utiliza la(s) técnicas(s) de conservación desde su introducción hasta hoy? (Si/No):______________ MCXXI
(Si "No": Preg. 13)

11. **Si "Si",** describa los principales elementos de la técnica y cuales han cambiado en comparación con el manejo original de la(s) técnica(s) actual(es).

Técnica de Conservación	**Cómo comenzó a hacer las técnicas?**	**Cómo hace las técnicas hoy?**
MCXXII	MCXXIII	MCXXIV
MCXXV	MCXXVI	MCXXVII
MCXXVIII	MCXXIX	MCXXX
MCXXXI	MCXXXII	MCXXXIII
MCXXXIV	MCXXXV	MCXXXVI

12. Por que se ha cambiado el manejo de la(s) técnica(s) mencionada(s)?

Técnica de Conservación	**Razón para el cambio en el Manejo de la Técnica**
MCXXXVII	MCXXXVIII
MCXXXIX	MCXL
MCXLI	MCXLII
MCXLIII	MCXLIV
MCXLV	MCXLVI

4. Arreglos Comunitarios

13. Dónde obtiene usted información sobre nuevas técnicas para el manejo de sus parcelas?______________________________ MCXLVII

14. Existen organizaciones/grupos de la comunidad que están involucrados en la promoción de técnicas de conservación? (Sí/No).__________ MCXLVIII
(Si "No": Final de la Encuesta)

15. **Si "Sí"**, Cuáles son?

1. ____ MCXLIX
2. ____ MCL
3. ____ MCLI

16. Que hacen estos grupos comunitarios?

Grupo coumunitario	Que hace?
MCLII	MCLIII
MCLIV	MCLV
MCLVI	MCLVII
MCLVIII	MCLIX
MCLX	MCLXI

17. Existen arreglos entre los agricultores o con la comunidad para dar ayuda con la introducción y/o el mantenimiento de las técnicas de conservación (cooperativas, directivas comunales, bancos comunales, etc.)? (Sí/No)_______ MCLXII *(Si "No": Final de la Encuesta)*

17. **Si "Sí".** Como funciona esta ayuda? MCLXIII

____
____
____
____
____
____
____
____

5. El No-Uso de Técnicas de Conservación

18. Conoce usted algunas técnicas de conservación? (Sí/No)__________ MCLXIV
(Si "No": Sigue pregundtadolc, insistiendo)

19. **Si "Sí"**, Cuáles conoce y de dónde?

Técnica conocida	**Dónde conoció la técnica por primera vez**
MCLXV	MCLXVI
MCLXVII	MCLXVIII
MCLXIX	MCLXX
MCLXXI	MCLXXII
MCLXXIII	MCLXXIV

20. Participó usted alguna vez en algún programa/curso de capacitación sobre las técnicas de conservación? (Sí/No).____________________ MCLXXV
(Si "No": Sigue preguntadole, insistiendo, Preg 22)

21. **Si "Sí"**, Con cual organización, cuándo y por cuánto tiempo?

Técnica conocida	**Institución**	**Tipo de Capacitación Duración días**	**Lugar**	**Año**
MCLXXVI	MCLXXVII	MCLXXVIII	MCLXXIX	MCLXXX
MCLXXXI	MCLXXXII	MCLXXXIII	MCLXXXIV	MCLXXXV
MCLXXXVI	MCLXXXVII	MCLXXXVIII	MCLXXXIX	MCXC
MCXCI	MCXCII	MCXCIII	MCXCIV	MCXCV
MCXCVI	MCXCVII	MCXCVIII	MCXCIX	MCC

22. Alguna vez ha usado usted una de estas técnicas? (Sí/No) ________ MCCI *(Si "Sí": Preg. 23, si "No": Preg. 24)*

23. **Si "Sí"**, Por que no sigue usando la(s) técnica(s) conocida(s)?

Técnica conocida	**Razón por el abandono de la técnica**
MCCII	MCCIII
MCCIV	MCCV
MCCVI	MCCVII
MCCVIII	MCCIX
MCCX	MCCXI

24. **Si "No"**, Porqué no usa la(s) técnica(s) conocida(s)?

Técnica conocida	**Razón por el no-uso de la técnica**
MCCXII	MCCXIII
MCCXIV	MCCXV
MCCXVI	MCCXVII
MCCXVIII	MCCXIX
MCCXX	MCCXXI

Observaciones del Encuestador

____

____

____

____

List of Figures

List of Tables